Integration of Process Planning and Scheduling

Science, Technology, and Management Series

Series Editor:
J. Paulo Davim
*Professor, Department of Mechanical Engineering,
University of Aveiro, Portugal*

This book series focuses on special volumes from conferences, workshops, and symposiums, as well as volumes on topics of current interests in all aspects of science, technology, and management. The series will discuss topics such as, mathematics, chemistry, physics, materials science, nanosciences, sustainability science, computational sciences, mechanical engineering, industrial engineering, manufacturing engineering, mechatronics engineering, electrical engineering, systems engineering, biomedical engineering, management sciences, economical science, human resource management, social sciences, engineering education, etc. The books will present principles, models techniques, methodologies, and applications of science, technology and management.

Advanced Mathematical Techniques in Engineering Sciences
Edited by Mangey Ram and J. Paulo Davim

Soft Computing Techniques for Engineering Optimization
Edited by Kaushik Kumar, Supriyo Roy, and J. Paulo Davim

Handbook of IOT and Big Data
Edited by Vijender Kumar Solanki, Vicente García Díaz, and J. Paulo Davim

Digital Manufacturing and Assembly Systems in Industry 4.0
Edited by Kaushik Kumar, Divya Zindani, and J. Paulo Davim

Optimization Using Evolutionary Algorithms and Metaheuristics
Edited by Kaushik Kumar and J. Paulo Davim

For more information about this series, please visit: https://www.crcpress.com/Science-Technology-and-Management/book-series/CRCSCITECMAN

Integration of Process Planning and Scheduling
Approaches and Algorithms

Edited by
Rakesh Kumar Phanden, Ajai Jain, and
J. Paulo Davim

CRC Press
Taylor & Francis Group
Boca Raton London New York

CRC Press is an imprint of the
Taylor & Francis Group, an **informa** business

CRC Press
Taylor & Francis Group
6000 Broken Sound Parkway NW, Suite 300
Boca Raton, FL 33487-2742

First issued in paperback 2021

© 2020 by Taylor & Francis Group, LLC
CRC Press is an imprint of Taylor & Francis Group, an Informa business

No claim to original U.S. Government works

ISBN-13: 978-0-367-03078-0 (hbk)
ISBN-13: 978-1-03-217686-4 (pbk)
DOI: 10.1201/9780429021305

Publisher's Note

The publisher has gone to great lengths to ensure the quality of this reprint but points out that some imperfections in the original copies may be apparent.

Library of Congress Cataloging-in-Publication Data
Names: Phanden, Rakesh Kumar, editor. | Jain, Ajai, editor. | Davim, J. Paulo, editor.
Title: Integration of process planning and scheduling : approaches and algorithms / edited by Rakesh Kumar Phanden, Ajai Jain, and J. Paulo Davim.
Description: Boca Raton : CRC Press, 2019. | Series: Science, technology, and management | Includes bibliographical references and index.
Summary: "Both process planning and scheduling are very important functions of manufacturing, which affects together the cost to manufacture a product and the time to deliver it. This book contains various approaches proposed by researchers, to integrate the process planning and scheduling functions of manufacturing under varying configurations of shops. It is useful for both beginners and advanced researchers to understand and formulate the Integration Process Planning and Scheduling (IPPS) problem effectively"— Provided by publisher.
Identifiers: LCCN 2019024675 (print) | LCCN 2019024676 (ebook) | ISBN 9780367030780 (hardback ; acid-free paper) | ISBN 9780429021305 (ebook)
Subjects: LCSH: Production scheduling.
Classification: LCC TS157 .I576 2019 (print) | LCC TS157 (ebook) | DDC 658.5/3—dc23
LC record available at https://lccn.loc.gov/2019024675
LC ebook record available at https://lccn.loc.gov/2019024676

Visit the Taylor & Francis Web site at
http://www.taylorandfrancis.com

and the CRC Press Web site at
http://www.crcpress.com

Contents

Preface...vii
Editors..ix
Contributors...xi

1. Integration of Process Planning and Scheduling: Introduction...........1
   Rakesh Kumar Phanden and Ajai Jain

2. Approaches to Integrate Process Planning and Scheduling 19
   Rakesh Kumar Phanden and Ajai Jain

3. A Case Study on Optimisation of Integrated Process Planning
   and Scheduling Functions Using Simulation-Based Genetic
   Algorithm and Heuristic for Makespan Performance
   Measurement 49
   Rakesh Kumar Phanden and Ajai Jain

4. An Approach to Integrated Process Planning and Scheduling
   Based on Variable Neighbourhood Search 95
   Oleh Sobeyko and Lars Mönch

5. Integration of Process Planning and Scheduling in an
   Energy-Efficient Flexible Job Shop: A Hybrid Moth Flame
   Evolutionary Algorithm 115
   *Vijaya Kumar Manupati, Subhash C. Bose, Rajeev Agrawal,
   Goran D. Putnik, and M.L.R. Varela*

6. Integration of Scheduling and Process Planning in Shop Floor:
   A Probability Model-Based Approach.. 135
   *R. Pérez-Rodríguez, A. Hernández-Aguirre, and
   S. Frausto-Hernández*

7. Integrated Process Planning and Scheduling Using Dynamic
   Approach 141
   Gideon Halevi and Rakesh Kumar Phanden

8. Due-Date Agreement in Integrated Process Planning and
   Scheduling Environment Using Common Meta-Heuristics 161
   Halil Ibrahim Demir and Rakesh Kumar Phanden

9. **Integration of Process Planning and Scheduling: An Approach Based on Ant Lion Optimisation Algorithm** 185
Milica Petrović and Zoran Miljković

10. **A Review on Testbed Problems for Integration of Process Planning and Scheduling** 207
Rakesh Kumar Phanden, Halil Ibrahim Demir, and Ajai Jain

Index 223

Preface

Process planning and production scheduling are the two most important functions of a manufacturing system. These functions strongly influence the profitability of manufacturing a product, the optimum utilization of resources, the delivery time of products, and many more factors. Conventionally, the process planning and scheduling activities are performed in a sequential manner, i.e., the process planning is done before scheduling. This approach involves many drawbacks such as unrealistic process plans without considering the shop floor status, fixed process plans without considering the alternative resources, invalid or suboptimal process plans and schedule due to delay in execution, optimized process plans and schedule with a single objective, etc. Thus, these problems can only be solved by the integration of both functions. The integration of process planning and scheduling (IPPS) can enhance the use of manufacturing flexibility in practice, it can improve the level of resource utilization and ultimately it can increase the overall profits from manufacturing. Consequently, this subject has been studied by several researchers over the last three to four decades in order to achieve the ultimate IPPS. This book is an attempt to summarize the approaches and algorithms applied for IPPS by different researchers across the world. It is worth mentioning that this is the only book which contains dedicated chapters on the introduction and fundamental approaches for IPPS, as well as a brief review on IPPS-tested problems. This book comprises many chapters authored by renowned researchers in IPPS research. Moreover, this book also discusses the role of IPPS in Industry 4.0 paradigm and energy-aware modelling for multi-objective IPPS.

Editors
Rakesh Kumar Phanden
Ajai Jain
J. Paulo Davim

MATLAB® is a registered trademark of The MathWorks, Inc. For product information,
   please contact:
   The MathWorks, Inc.
   3 Apple Hill Drive
   Natick, MA 01760-2098 USA
   Tel: 508-647-7000
   Fax: 508-647-7001
   E-mail: info@mathworks.com
   Web: www.mathworks.com

Editors

Rakesh Kumar Phanden completed his graduation in Mechanical Engineering from the U.P. Technical University, Lucknow, India and post-graduation in Integrated Product Design and Manufacturing from the Department of Mechanical Engineering, G.J.U. of Science and Technology, Hisar, India. He completed his Ph.D. in Mechanical Engineering from the National Institute of Technology, Kurukshetra, India in 2013. He is currently working in the Department of Mechanical Engineering at the Amity University, Uttar Pradesh, India. He has 12 years of teaching experience at private and government institutes and universities. He has contributed more than 30 papers at the national/international levels. His current areas of interest include manufacturing systems, production scheduling, integration of process planning and scheduling and product design and manufacturing.

Ajai Jain completed his Ph.D. from Kurukshetra University, India in 2003. He did his Masters of Engineering from the Department of Production & Industrial Systems Engineering, University of Roorkee, India in 1993. Dr. Jain has secured masters and bachelor with honors. He is working as a professor in the Mechanical Engineering Department of National Institute of Technology, Kurukshetra, India. He has 25 years of experience teaching undergraduate and postgraduate courses in mechanical engineering. He is a Fellow member of the Institution of Engineers (India) and Life Member of ISTE. He has worked as a Principal Investigator in a project on Integration of Process Planning and Scheduling sponsored by the Science and Engineering Research Council, Department of Science and Technology, New Delhi, Govt. of India. He has delivered many lectures in short-term courses. Dr. Jain has supervised 7+ Ph.D. candidates. His research area includes design and operation of manufacturing systems, process planning, production scheduling, integration of process planning and scheduling, reconfigurable manufacturing systems, advanced machining process, WEDM, computer-aided manufacturing, computer-integrated manufacturing. He has published many research articles in the journals of international repute. His SCOPUS h-index is 13+.

J. Paulo Davim received his Ph.D. degree in Mechanical Engineering in 1997, M.Sc. degree in Mechanical Engineering (materials *and* manufacturing processes) in 1991, Mechanical Engineering degree (5 years) in 1986, from the University of Porto (FEUP), the Aggregate title (Full Habilitation) from the University of Coimbra in 2005 and the D.Sc. from London Metropolitan University in 2013. He is a senior chartered engineer by the Portuguese Institution of Engineers with an MBA and Specialist title in Engineering and Industrial Management. He is also Eur Ing by FEANI-Brussels and Fellow (FIET) by IET-London. Currently, he is Professor at the Department of Mechanical Engineering of the University of Aveiro, Portugal. He has more than 30 years of teaching and research experience in Manufacturing, Materials, Mechanical and Industrial Engineering, with special emphasis in Machining & Tribology. He also has interest in Management, Engineering Education and Higher Education for Sustainability. He has guided large numbers of postdoc, Ph.D. and master's students, and has coordinated and participated in several financed research projects. He has received several scientific awards. He has worked as evaluator of projects for ERC-European Research Council and other international research agencies as well as examiner of Ph.D. thesis for many universities in different countries. He is the editor in chief of several international journals, guest editor of journals, book editor, book series editor, and scientific advisory for many international journals and conferences. Presently, he is an editorial board member of 30 international journals and acts as reviewer for more than 100 prestigious *Web of Science* journals. In addition, he has also published as editor (and co-editor) more than 100 books and as author (and co-author) more than 10 books, 80 book chapters, and 400 articles in journals and conferences (more than 250 articles in journals indexed in *Web of Science* core collection/h-index 50+/7500+ citations, SCOPUS/h-index 56+/10500+ citations, Google Scholar/h-index 71+/16500+).

Contributors

Rajeev Agrawal
Department of Mechanical
  Engineering
Malaviya National Institute of
  Technology Jaipur
Jaipur, India

Subhash C. Bose
Department of Mechanical
  Engineering
National Institute of Technology
  Warangal
Warangal, India

Halil Ibrahim Demir
Department of Industrial
  Engineering
Sakarya University & Artificial
  Intelligence Systems Application
  and Research Centre
Sakarya, Turkey

S. Frausto-Hernández
Department of Chemical Engineering
Technological Institute of
  Aguascalientes
Aguascalientes, México

Gideon Halevi
Retired – Director of CAD/CAM
  R&D Center at IMI Corporation
  Technion
Haifa, Israel

A. Hernández-Aguirre
Department of Computer Science
Center for Mathematics Research
  México
Guanajuato, México

Ajai Jain
Department of Mechanical
  Engineering
National Institute of Technology
  Kurukshetra
Kurukshetra, India

Vijaya Kumar Manupati
Department of Mechanical
  Engineering
National Institute of Technology
  Warangal
Warangal, India

Zoran Miljković
Department of Production
  Engineering
University of Belgrade
Kraljice Marije, Serbia

Lars Mönch
Department of Mathematics and
  Computer Science
University of Hagen
Hagen, Germany

Milica Petrović
Department of Production
  Engineering
University of Belgrade
Kraljice Marije, Serbia

Rakesh Kumar Phanden
Department of Mechanical
  Engineering
Amity University Uttar Pradesh
Noida, India

Goran D. Putnik
Department of Production and
   Systems
University of Minho
Guimarães, Portugal

R. Pérez-Rodríguez
National Council for Science and
   Technology
CONACYT - Center for Mathematics
   Research México
Guanajuato, Mexico

Oleh Sobeyko
Department of Mathematics and
   Computer Science
University of Hagen
Hagen, Germany

M.L.R. Varela
Department of Production and
   Systems
University of Minho
Guimarães, Portugal

1

Integration of Process Planning and Scheduling: Introduction

Rakesh Kumar Phanden

Amity University Uttar Pradesh

Ajai Jain

National Institute of Technology Kurukshetra

CONTENTS

1.1   Introduction 2
1.2   Process Planning 2
1.3   Classification of Process Planning 4
    1.3.1   Manual Process Planning 4
    1.3.2   Computer-aided Process Planning (CAPP) 4
        1.3.2.1   Variant CAPP Approach 4
        1.3.2.2   Generative CAPP Approach 5
        1.3.2.3   Semi-generative CAPP Approach 5
1.4   Master Production Scheduling (MPS) 6
1.5   Detailed Scheduling 6
    1.5.1   Completion Time 8
    1.5.2   Makespan 8
    1.5.3   Flow Time 8
    1.5.4   Lateness 9
    1.5.5   Tardiness 9
    1.5.6   Number of Tardy and Early Jobs 9
    1.5.7   Throughput 9
1.6   Classification of Scheduling Approaches 9
1.7   Job Shop Scheduling 10
1.8   IPPS 11
1.9   IPPS in Industry 4.0 Paradigm 13
1.10 The Takeaway 14
References 15

1.1 Introduction

Manufacturing companies are striving for effective allocation of human resources to equipment/machines, as well as effective allocation of equipment/machines to raw material (product to be manufactured), in order to achieve better business results by enhancing the shop floor efficiency. There are different manufacturing functions which play an important role in enhancing the business output such as process planning, estimating and costing, scheduling, dispatching, inspection and expediting materials control, loading/unloading, etc. However, process planning, as well as scheduling, are the utmost commanding and critical functions that directly affect the performance of a production system. Also, *Integration of Process Planning and Scheduling* (IPPS) is essential to break down the wall between the two departments by increasing the pace of information exchange. This chapter is presenting the basics of process planning, production scheduling, IPPS, configurations of the manufacturing environment and its performance measures, as well the role of IPPS in the Industry 4.0 paradigm.

1.2 Process Planning

Process planning is performed in both discrete and process industries. In general, process planning can be divided into machining planning and assembly planning. In the former, machining instructions are determined for each part on an individual machining centre, whereas in assembly planning, as the name represents, only assembly instructions are determined to form a product. Basically, machining planning has been simplified as process planning or manufacturing process planning. It is important to understand the various terms like *process*, *operation*, and *operation sequencing* used in process planning. *Process* means a continuous procedure in which one or more parts are machined (treated or processed) on one or more machines by one or more operators, for instance, the rough drilling process of a hole, finish drilling process of the hole by a drilling machine, grinding process on a grinder wheel, etc. *Operation* is the part of a process in which the machining surface, machine tool, and machining conditions (feed, speed, depth of cut) remain unchanged. Operation is one small section of a process. When all operations of a process are arranged in the desired order, the arrangement is called *operation sequencing*. According to Society of Manufacturing Engineers (SME), process planning is defined as a *systematic determination of the methods by which a product is manufactured economically and competitively*. It connects design and manufacturing stages and involves a great amount of decision-making among various engineering activities. Process planning follows the procedural steps which are initialised with the recognition of product design specifications (data), trailed by the selection of manufacturing processes, machine tools, fixtures and datum surfaces, the sequence of operations, inspection procedures, determination of tolerances and cutting conditions, and the calculation of total processing time. These detailed instructions to transform the raw material into the finished product are documented on the process sheet (also termed as *operation sheet*, *planning sheet* or *route sheet* or *route plan* or *part program* or *process plan*). Different types of data related to product design, material, equipment, and quality are the input to process planning function and process plan is the output, as presented in Figure 1.1. Thus, process planning is an engineering activity and process plan is the outcome of it in the form of documentation of the manufacturing instructions.

The process planning influences the cost of production as well as the time to market. A process planner strives to produce the optimal process plan by considering each production activity and parameters to satisfy the desired criteria of production cost, time, and quality. Therefore, he must possess rich manufacturing knowledge and experience to comprehend the engineering drawing, bills of material, and machining equipment. Thus, process planning plays a key role in gaining competitive advantage.

The process planning can be categorised as "macro process planning" and "micro process planning". The former includes the analysis of availability and accessibility of machining processes and manufacturing features, selection of machine tool and process, selection of operation sequences and setup planning, whereas the micro process planning involves the determination of cutting conditions, process optimisation, analysis of machining performance, optimisation of tool path, and trade-off amid part-process-fixture design.

The function of process planning can be extended up to the customer end if it involves transportation, packaging, metrology, and export activities of the final product. This inclusion makes the process planning function a much more

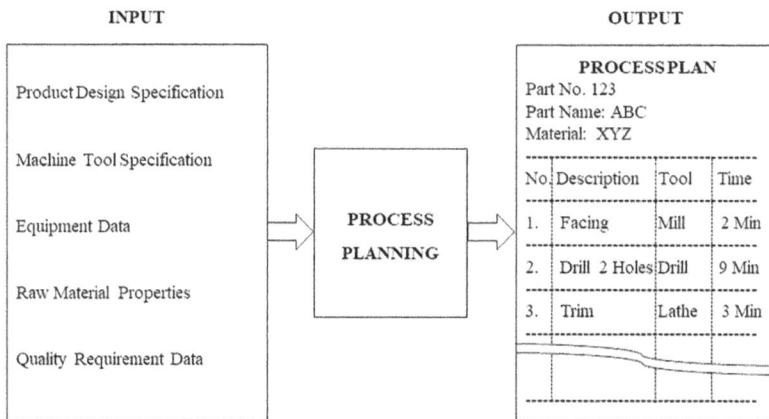

FIGURE 1.1
Basic model of process planning.

complex task rather than being just a manufacturing recipe. In general, these are routine activities and are performed by creative and skilled experts. Hence, the process planning is a team activity that involves specialists from different domains, such as jigs and fixture designer, tool designer, cutting condition experts, and process planner, etc. A good process plan is the output of multiparty cooperation and effective information exchange among dedicated subsystems.

1.3 Classification of Process Planning

The process planning can be categorised as *manual process planning* and *computer-aided process planning*.

1.3.1 Manual Process Planning

Basically, manual process planning has been perceived as a traditional and a workbook approach. The former approach involves the study of the blueprint (of part design specification) and manual of machine tool and equipment to determine the suitable processes, machine, jig fixture, material and tools for the selected part, to form the routing sheet. In the workbook approach, process planner identifies the required manufacturing processes and uses the standardised sequence of operations from a canned workbook (for a class of product) to form the operation sheet. The manual process planning approaches need a skilled team of the planner. Also, it is a time-consuming method and lacks consistency to dynamically update the process plan for a fresh product.

1.3.2 Computer-aided Process Planning (CAPP)

CAPP can be defined as *automation of process planning through the computer support and cutting-edge planning approaches [based on group technology (GT) and exact mathematical principles] to set up the fast information exchange amid design and manufacturing activities and to generate the viable process plans readily.* It works as a channel amid the manufacturing and design departments in a computer-integrated manufacturing environment. CAPP can generate the feasible process plans considering various factors like manufacturing lead time, cost and volume of production, availability of machines and equipment, alternative material, and process routing, etc. In the last two decades, several CAPP systems have been established on the bases of *variant approach, generative approach*, and *semi-generative approach* to perform the process planning activity.

1.3.2.1 Variant CAPP Approach

Here, an existing process plan is edited to make it suitable for a fresh product. It works on the idea of GT. Basically, it follows two stages, namely, the "preparatory stage" and the "production stage". In the former, the coding scheme and part family are designed, which is followed by the generation of standard process plans. At the production stage, the standard process plan is retrieved and modified to form a generic process plan for a new part. In fact, the part families and the classification schemes of GT are formed according to the similarities in terms of design features and manufacturing processes used to machine the parts.

Wolfe (1985) developed MAYCAPP variant CAPP system using FORTRAN programming language and it works based on DCCLASS coding characteristics. Tulkoff (1987) built CUTPLAN for rotational and prismatic parts for process determination only. CAPP-I, TOJICAP, WICAPP, CAM-I CAPP, MIPLAN, MITURN, MIAPP, ACUDATA SAPT, and many more are the variant CAPP systems developed based on GT. All the variant CAPP systems lack the ability of generating the process plan for an exclusively new product. Also, this approach requires human expertise to intervene for retrieving and editing of the existing standard process plan, which leads to poor quality of the process plan.

1.3.2.2 Generative CAPP Approach

This is a fully automatic approach to generate process plans for a new part automatically by gathering process information from scratch. The process-planning knowledge is applied to a generative CAPP system using process logics like decision trees and tables and constraint-based or rule-based logic to select the manufacturing process by matching the process capability with the design requirement of the part. Also, this CAPP system is functional to select process parameters, machine tools, and other production parameters automatically. Basically, the generative CAPP approach is consistent, automatic, and fast to generate the process plans for any size of the product mix. Numerous generative CAPP systems have been developed, namely APPAS, AUTOPLAN, GENPLAN, TIPPS, AUTAP, TURBO-CAPP, TOM, ICAPP, XCUT, XPLANE, CMPP, and ICAPP. A survey of generative CAPP systems is given by Gupta and Ghosh (1989).

1.3.2.3 Semi-generative CAPP Approach

It is a combination of both the variant and generative approaches in which the process plan is generated using generative approach and the same is improved (through human intervention to examine for errors and deviations from standard process plan) before the commencement of production, if required. GENPLAN, ACAPS, MICRO-GEPPS, BITCAPP, AC/PLAN, AUSPLAN, CORECAPP, CRUNCH, DAPP, DCLASS, POPULAR, XPS-1, and ZCAPPS are the semi-generative CAPP systems developed by different researchers and agencies across the world.

Niebel (1965) presented the first idea to do process planning using computer support due to its speed and consistency. Thereafter, many researchers have made plentiful efforts to develop and implement an effective CAPP system. Zhang and Alting (1994) have presented several CAPP systems proposed and developed by researchers and companies across the world. Despite these efforts, the existing CAPP systems are not suitable to meet the expectations of industries in order to implement in a real manufacturing system. These CAPP systems are derived from the living status of the shop floor during the process plan generation. Various shop floor conditions like capacity and availability (workload) of machines, bottlenecks and breakdowns, product mix, production scheduling, and other controlling aspects always play a vital role in achieving the desired performance of a manufacturing system. Therefore, these issues should be considered while performing process planning.

1.4 Master Production Scheduling (MPS)

Traditionally, MPS is a decision-making process which generates an anticipated overall schedule (dictating the varieties, quantities, and dates to manufacture end products) based on output data of production capacity, demand forecast, backlog, aggregate production plan, availability and flow of materials, inventory levels, etc. MPS acts as a bridge amid marketing and manufacturing activities. It is a report of production, not a forecast. Although the sales forecast is an input to generate MPS. It is also used to find the prospect of revenue generation of production orders received. However, the discussion on MPS is beyond the scope of this chapter.

1.5 Detailed Scheduling

It focuses on a shorter time horizon than MPS and is the most complex and precise shop floor level scheduling method that plays a key role in enhancing the productivity of manufacturing. In general, the scheduling (also termed as "detailed scheduling" or "production scheduling" or "manufacturing scheduling") is a decision-making process of assigning the task to resources over a time-period based on detailed resources availability and usages. Specifically, in a manufacturing system, it allocates machines to operations and/or operations to machines dictated in the process plan to achieve the desired objectives (performance measures) like throughput, flow time, completion time, lateness, or tardiness of products. Basically, the scheduling performance measures are varying from business to business based on due-date of delivery, inventory levels, and completion time of products. Various regularly used performance measures of scheduling are discussed below. Furthermore, Tables 1.1 and 1.2 are presenting formulations and problem-types of scheduling objectives,

TABLE 1.1

Scheduling Objective and Its Formulation

	Completion Time	Flow Time	Lateness	Tardiness	Earliness
Total	$\displaystyle\sum_{i=1}^{n} C_i$	$\displaystyle\sum_{i=1}^{n}(C_i - R_i)$	$\displaystyle\sum_{i=1}^{n}(C_i - D_i)$	$\displaystyle\sum_{i=1}^{n}\max\{0,(C_i - D_i)\}$	$\displaystyle\sum_{i=1}^{n}(D_i - C_i)$
Total Weighted	$\displaystyle\sum_{i=1}^{n} W_i\, C_i$	$\displaystyle\sum_{i=1}^{n} W_i(C_i - R_i)$	$\displaystyle\sum_{i=1}^{n} W_i(C_i - D_i)$	$\displaystyle\sum_{i=1}^{n} W_i\big[\max\{0,(C_i - D_i)\}\big]$	$\displaystyle\sum_{i=1}^{n} W_i(D_i - C_i)$
Average	$\displaystyle\frac{1}{n}\sum_{i=1}^{n} C_i$	$\displaystyle\frac{1}{n}\sum_{i=1}^{n}(C_i - R_i)$	$\displaystyle\frac{1}{n}\sum_{i=1}^{n}(C_i - D_i)$	$\displaystyle\frac{1}{n}\sum_{i=1}^{n}\max\{0,(C_i - D_i)\}$	$\displaystyle\frac{1}{n}\sum_{i=1}^{n}(D_i - C_i)$
Average Weighted	$\displaystyle\frac{1}{n}\sum_{i=1}^{n} W_i\, C$	$\displaystyle\frac{1}{n}\sum_{i=1}^{n} W_i(C_i - R_i)$	$\displaystyle\frac{1}{n}\sum_{i=1}^{n} W_i(C_i - D_i)$	$\displaystyle\frac{1}{n}\sum_{i=1}^{n} W_i\big[\max\{0,(C_i - D_i)\}\big]$	$\displaystyle\frac{1}{n}\sum_{i=1}^{n} W_i(D_i - C_i)$
Maximum	$\max$ $(C_1, C_2,\ldots, C_n)$	$\max(C_1 - R_1),$ $(C_2 - R_2),\ldots,$ $(C_n - R_n)$	$\max(C_1 - D_1),$ $(C_2 - D_2),\ldots,$ $(C_n - D_n)$	$\max\{0,(C_1 - D_1)\},$ $(C_2 - D_2),\ldots,(C_n - D_n)\}$	$\max(D_1 - C_1),$ $(D_2 - C_2),\ldots,$ $(D_n - C_n)$

C_i, completion time of job I; W_i, relative weight of job I; R_i, release date of job I; D_i, due date of job i.

TABLE 1.2

Scheduling Objective and Problem Type

	Completion Time	Flow Time	Lateness	Tardiness	Earliness
Total	MIN-SUM	MIN-SUM	MIN-SUM	MIN-SUM	MAX-SUM
Total weighted	MIN-SUM	MIN-SUM	MIN-SUM	MIN-SUM	MAX-SUM
Average	MIN-SUM	MIN-SUM	MIN-SUM	MIN-SUM	MAX-SUM
Average weighted	MIN-SUM	MIN-SUM	MIN-SUM	MIN-SUM	MAX-SUM
Maximum	MIN-MAX	MIN-MAX	MIN-MAX	MIN-MAX	MAX-MAX

MAX-MAX, maximise maximum of quantity; MIN-MAX, minimise maximum of quantity; MIN-SUM, minimise sum of quantity; MAX-SUM, maximise sum of quantity.

TABLE 1.3

Decision-making Criterion Versus Scheduling Objectives (Jain and Meeran, 1999)

Decision-making Goal	Scheduling Objective
Effective utilisation of resources	Makespan
Fast response to demand and minimises work-in-process	Flow time
Strict following of the given deadline	Tardiness, lateness, earliness, number of tardy and early jobs

respectively. Table 1.3 shows the decision-making goal versus frequently used scheduling objectives.

1.5.1 Completion Time

The time consumed for the processing (set of operations) of a job is called the completion time of that job.

1.5.2 Makespan

It is the completion time of processing of the last job in a production (batch) order. In other words, it is the maximum completion time consumed to process a production order.

1.5.3 Flow Time

It is the time spent by a job in the shop (i.e., shop time of a job). In other words, the time period within the job released time and the processing accomplishment time of a job is called flow time.

1.5.4 Lateness

Lateness is the time difference between the completion time and delivery due date of a job. Besides, the opposite of lateness is termed as an earliness performance measure. Thus, if the lateness is negative then the earliness will be positive and vice-versa.

1.5.5 Tardiness

It is the same as lateness but carries only positive values. If a job was accomplished ahead of its due date, its lateness becomes a negative value and tardiness becomes zero.

1.5.6 Number of Tardy and Early Jobs

A number of tardy and early jobs are the measurements of the number of jobs in a production order which are processed (accomplished) later, and before the prescribed due dates of jobs, respectively.

1.5.7 Throughput

It is defined as "the number of parts produced in a unit time". In general, it is a measurement of the output rate of a machine (in a manufacturing system) having the lowest production volume compared to the others, due to bottlenecks.

The formulation of completion time, flow time, lateness, and tardiness for their components of the total, total weight, average, and average weight are in the category of minimising the sum of a quantity. In earliness scheduling objective, the problem formulation is for the maximisation of the sum of a quantity. However, the formulation for maximisation component of objective scheduling (given in Table 1.1) is the problem of minimising the maximum of a quantity. In earliness, the same is the problem of maximisation of the maximum of a quantity.

1.6 Classification of Scheduling Approaches

In general, the approach for scheduling is dependent upon the types of manufacturing systems involved. Various real-world shop configurations are existing, like job shops, open shops, flow shops, single-machine shops, mixed shops, etc. In brief, Table 1.4 provides a description of the types of shop configurations. There are two ways to do scheduling based on time,

TABLE 1.4

Types of Shop Configurations (Pinedo, 2008)

Types	Configurations
Single machine shop	It is a type of shop which possesses only one machine and all the jobs are processed on that machine.
Flow shop	The flow shop consists of "m" number of machines installed in a sequence in which each job is processed as per its sequence of operation. In other words, it is an arrangement of machines as per the sequence of operations of jobs.
Job shop	It is configured with "m" number of machines to process "n" types (or batches) of jobs having a dissimilar sequence of operations.
Open shop	The open shop consists of "m" number of machines in which jobs are routed unrestrictedly (openly) through any machine without ensuring the stated sequence of operation. In this configuration, each job (of production order) may have a different route.
Mixed shop	In this shop, "m" number of machines are arranged to process "n" number of jobs with fixed, as well as a varying route (sequence of operations).

i.e., (a) offline scheduling, (b) online scheduling. Offline scheduling is also called predictive scheduling. This type of scheduling plays a cardinal role in strategic planning. It works offline based on on-hand information. Offline scheduling narrates allocation of resources available on the shop floor. Also, it assists in decision-making for various external functions like preventive maintenance, material procurement, etc. Owing to the many unexpected external and internal disturbances, like rush order entrance and machine breakdown, etc., the existing (initial) schedule may become impracticable and it needs to be brought up to date. Hence, this type of scheduling approach is called re-scheduling. In the online scheduling process, the schedule is generated during the production process. It follows the up-to-date status of shop floor and generates the schedule based on the real-time approach.

1.7 Job Shop Scheduling

Most small to medium size manufacturing companies are working on the job shops environment and it is suitable to meet individual customer needs (Chryssolouris, 2005). Therefore, the job shop is an important type of configuration that delivers credible understanding into the solution of the production schedule as well as other common scheduling problems faced in realistic environments. Job shop scheduling is a non-polynomial hard combinatorial optimisation problem (Davis, 1985). A job shop is configured with "m" number of machines to process "n" types (or batches) of jobs having a dissimilar sequence of operations. Job shop scheduling problem deals with the allocation of machines to jobs or vice-versa, to achieve the desired objectives such as minimisation of makespan, tardiness, mean flow time, etc. Also, it is subjected to various constraints and assumptions. In a job shop, the flow of work changes the travel direction often and a machine could be regarded by the output and input flows of work (Phanden et al., 2012). The job shop scheduling problem has been solved by mixed integer programming, branch and bound method, disjunctive graph method, shifting bottlenecks algorithms, and dispatching rules. Different evolutionary algorithms such as tabu search, simulated annealing, ant colony optimisation, neighborhood search, particle swarm optimisation, cuckoo search, genetic algorithm (GA), etc., have been applied to optimise the schedule for a job shop manufacturing environment (Jain and Meeran, 1999; Phanden et al., 2018). Several meta-heuristic algorithms to optimise the job shop scheduling problem have been discussed in Jain and Meeran (1999). They revealed that among these approaches, the growing concentration is in applying a GA. Also, GA is a well-known optimisation algorithm for the class of combinatorial problems. It embraces parallelism, constructive redundancy as well as past information storage capability to apply on enormously parallel architectures (Wang and Zheng, 2002).

It has been observed that a job shop is solved for static and dynamic scheduling environment by various researchers. Although there is still a gap between scheduling theory and practice. Therefore, the results of theoretical testing cannot be implemented in real setups directly. This is because most of the scheduling theories have considered job shop scheduling problem in isolation from higher-level functions, like process planning, that play a key role in the management of the manufacturing system (Gupta, 2002). Thus, the interaction between higher-level decision-making functions of management and lower-level scheduling function is a must to improve the performance of the manufacturing system and to implement the scheduling plan in real practice.

1.8 IPPS

Both, the process planning and scheduling functions play an important role in all types of manufacturing systems. These functions have an influence on the profit share production and on-time delivery of the product, as well as on the optimum utilisation of manufacturing resources (Gindy et al., 1999; Baykasoğlu and Özbakır, 2009; Zhang and Wong, 2016). As discussed earlier, the process planning is the *systematic determination of methods by which a product is to be manufactured economically and competitively*. The prime objective of process planning activity is to produce the process plans that can lay down the raw material and constituents required to produce a product, as well as a combination of operations on the part and its sequences to process on machines, in order to convert the raw material into the final product. Hence, the output of process planning activity is the information needed for manufacturing processes that includes identification of machine tool, jig and fixtures, operation setups, etc. On the other hand, production scheduling acts on assigning a particular job to a particular machine in order to achieve the desired performance of the manufacturing system. Scheduling is restricted by the instructions issued on process sequencing in process plans, as well as by the time-bound serviceability of manufacturing resources on the shop floor. Hence, both functions embrace the assignment of resources and perform in conjunction. Traditionally, process planning and scheduling tasks are performed in two different sequential steps in which production scheduling is carried out distinctly after process planning. This idea of execution is based on the theory of dividing the work into separate and small amount of duties and responsibilities in order to attain the suitability for mass production and suboptimisation methods (Phanden et al., 2011). However, the manufacturing scenario is drastically changing day by day to achieve various objectives, such as reduction of manufacturing lead time, production of a vast variety of products, meeting standards of product quality and customers' needs at a viable cost, customised (make-to-order) demands of customers, as well as to reduce energy consumption and pollutant emissions (Gindy et al., 1999; Baykasoğlu and Özbakır, 2009; Zhang and Wong, 2016). It is difficult to achieve these objectives in the traditional manufacturing systems because of various problems in the existing segregated functioning of process planning and scheduling, for instance:

- A process planner works on the assumption that the shop floor is in idle condition and resources have unconstrained abilities to process the part always. Thus, process planner plans for the most recommended alternative resources. This leads to the process planner favouring to select the desirable resources repeatedly. Moreover, the resources are never always available on the shop floor. Therefore, an unrealistic process plan will get generated that may not be readily executed on the shop floor.
- In a traditional manufacturing environment, the fixed process plans confine the production schedule to a specific machine and operation. Consequently, the probable options of production schedule via alternative machines are disregarded unintentionally.
- Even though at the beginning, the updated status of shop floor may have been considered while performing process planning, the constraints taken during process planning phase may change due to delays in execution time of scheduling and planning phases. Hence, an invalid schedule and suboptimal process plan are generated.
- Process planning and scheduling functions are solved for a single objective performance measure in order to find the suboptimal

results. Although, in an actual production setting, there are more than one objectives involved (Sugimura et al., 2001; Baykasoğlu and Özbakır, 2009).

Hence, the problems associated with the conventional process planning and scheduling can only be solved through the integration of both functions. The integrated system can perform better than the conventional manufacturing and it is capable of using the enhanced level of manufacturing flexibility, resulting in better utilisation of resources, well-timed delivery of products, and increased overall profitability from production. It can form the genuine process plans which can be implemented on the shop floor without modifications (Saygin and Kilic, 1999; Gindy et al., 1999; Lee and Kim, 2001; Kumar and Rajotia, 2002, 2003; Guo et al., 2009; Phanden et al., 2011; Liu et al., 2018). Hence, the IPPS is important in order to attain the ultimate integrated manufacturing system.

1.9 IPPS in Industry 4.0 Paradigm

Industry 4.0 is the next industrial revolution in which the building of smart factories has been started. In other words, conservative manufacturing approaches are changing into smart manufacturing. Machines are learning to understand processes, interacting with the environment, and smartly adopting their behaviour. Artificial intelligence (AI) and big data technologies are making machines more adaptable and smarter than before. Particularly, machine learning (ML) algorithms in the arena of AI, have facilitated the innovations in the industrial sector, which are yielding opportunities to further improve decision-making capability of manufacturing systems. ML algorithms learn from available data, experiences, and find out the most suitable way to accomplish a complex task (Bajic et al., 2018).

In fact, Industry 4.0 paradigm describes the manufacturing via Cyber Physical System (CPS). It is an embedded system in which the data of physical processes are captured through sensors and put into action over the digital networks (internet) and monitored by algorithms. If CPS is applied in production, it is termed as Cyber Physical Production System (CPPS) (Kagermann et al., 2013). It is composed of various independent, interconnected elements and subsystems over the complete manufacturing system, and embrace the characteristics of responsiveness, intelligence, and connectedness (Monostori et al., 2016). Moreover, the domain of Industry 4.0 is not specified and there is a lack of clear classifications of technologies and their features (Kagermann et al., 2013; Lu, 2017; Hermann et al., 2016). Meissner and Aurich (2019) identified that the "connectedness", "smart machines and products" (SM&P), "decentralisation" (DC), "cybersecurity" (CS), and "big data" (BD) are the most correlated and core elements from the manufacturing viewpoint. They stated that CS has no impact on both process planning and scheduling functions. Process planning does not get impacted by SM&P directly, however, it is desirable to perform distributed control because it aids in exchanging the information and decision-making for successive operations. Similarly, "connectedness" does not impact process planning function directly, but it helps in information exchanging process. The impact of these core elements can be identified at a technical level or methodical level. "Connectedness" and SM&P influence the process planning and scheduling at the technical level. In addition, other elements influence both the process planning and scheduling functions at the methodical level. Thus, it identifies dual impacts on process planning and scheduling functions. Primarily, DC indicates that jobs can find their sequence of operations and processes themselves, so they have different production path. Therefore, in order to inculcate this flexibility for jobs, the multiple (alternative) process plans (MPP) are a must to generate as they find the prospective paths. However, these process plans yield poor output as compared to the ideal process plan, but it helps to add flexibility in the system. Therefore, MPP must be forecasted for any CPPS-enabled IPPS approach. In addition, the combined benefits from the big data management and access of real-time data take the process planning function to the next level. Usually, the information flows in a unidirectional manner, from process planning to scheduling. The system could be updated if the information flows from scheduling to process planning but, the information may get outdated by the time it reaches the process planning department. The global accessibility of real-time information yields novel evidence to perform accurate process planning.

1.10 The Takeaway

Process planning is a *systematic determination of the methods by which a product is manufactured economically and competitively*. The manual process planning is a time-consuming method and lacks consistency to dynamically update the process plan for a new part. Variant CAPP approaches work on the GT, in which the existing process plan is edited to make it suitable for a fresh product. Generative CAPP is a fully automatic approach to generate process plans for a new part automatically by gathering process information from scratch. A job shop is an important type of configuration that delivers a good understanding of the solution of the production scheduling activity as well as other common scheduling problems faced in realistic environments. Most small to medium size manufacturing companies are working on the job shops environment and it is suitable to meet individual customer needs. It is configured with a number of machines to process "*n*" types (or batches) of jobs having a dissimilar sequence of operations. Thus, process planning and scheduling are the two most important functions of a manufacturing system. Both are interrelated and involve the assignment of resources. Therefore, the integration of these two functions is a must to achieve an integrated manufacturing environment. The integrated system can perform better than the conventional manufacturing and it is capable of using the enhanced level of manufacturing flexibility, resulting in better utilisation of resources, on-time delivery of products, and increased overall profitability from production. It can form the genuine process plans which can be implemented on the shop floor without modifications. The core elements of Industry 4.0 such as "connectedness", "smart machines and products", "decentralisation", and "big data" have been recognised as an important technology to achieve IPPS.

References

Bajic, B., Ignjatić, J., Suzic, N. and Stevanov, B. (2018). Machine Learning Techniques for Smart Manufacturing: Applications and Challenges in Industry 4.0. *In the Proceedings of TEAM 2018, 9th International Scientific and Expert Conference, under the auspices of the International TEAM Society*, October 10–12, 2018, University of Novi Sad, Serbia. pp. 29–38.

Baykasoğlu, A. and Özbakır, L. (2009). A grammatical optimization approach for integrated process planning and scheduling. *Journal of Intelligent Manufacturing*, 20(2), 211–221.

Chryssolouris, G. (2005). *Manufacturing Systems—Theory and Practice*, 2nd edition, Springer-Verlag, New York.

Davis, L. (1985). Job Shop Scheduling with Genetic Algorithms. *In the Proceedings of the 1st International Conference on Genetic Algorithms*, Pittsburgh, PA, In J. Grefenstette (ed.), pp. 136–140.

Gindy, N., Saad, S. and Yue, Y. (1999). Manufacturing responsiveness through integrated process planning and scheduling. *International Journal of Production Research*, 37(11), 2399–2418.

Guo, Y.W., Li, W.D., Mileham, A.R. and Owen, G.W. (2009). Applications of particle swarm optimisation in integrated process planning and scheduling. *Robotics and Computer-Integrated Manufacturing*, 25(2), 280–288.

Gupta, J.N.D. (2002). An excursion in scheduling theory: An overview of scheduling research in the twentieth century. *Production Planning & Control*, 2, 105–116.

Gupta, T. and Ghosh, B.K. (1989). A survey of expert systems in manufacturing and process planning. *Computers in Industry*, 11(2), 195–204.

Hermann, M., Pentek, T. and Otto, B (2016). Design Principles for Industrie 4.0 Scenarios. *In the Proceedings of the 49th Hawaii International Conference on System Sciences*, Koloa, HI, pp. 3928–3937.

Jain, A.S. and Meeran, S. (1999). A state-of-the-art review of job-shop scheduling techniques. *European Journal of Operations Research*, 113, 390–434.

Kagermann, H., Wahlster, W. and Helbig, J. (2013). Recommendations for implementing the strategic initiative Industrie 4.0. http://thuvienso.dastic.vn:801/dspace/handle/TTKHCNDaNang_123456789/357

Kumar, M. and Rajotia, S. (2003). Integration of scheduling with computer aided process planning. *Journal of Materials Processing Technology*, 138, 297–300.

Kumar, M. and Rajotia, S. (2006). Integration of process planning and scheduling in a job shop environment. *The International Journal of Advanced Manufacturing Technology*, 28(1–2), 109–116.

Lee, H. and Kim, S. (2001). Integration of process planning and scheduling using simulation based genetic algorithms. *The International Journal of Advanced Manufacturing Technology*, 18, 586–590.

Liu, M., Yi, S., and Wen, P. (2018). Quantum-inspired hybrid algorithm for integrated process planning and scheduling. *Proceedings of the Institution of Mechanical Engineers, Part B: Journal of Engineering Manufacture*, 232(6), 1105–1122.

Lu, Y. (2017). Industry 4.0: A survey on technologies, applications and open research issues, *Journal of Industrial Information Integration*, 1–10.

Meissner, H. and Aurich, J.C. (2019). Implications of cyber-physical production systems on integrated process planning and scheduling, *Procedia Manufacturing*, 28, 167–173

Monostori, L., Kádár, B., Bauernhansl, T., Kondoh, S., Kumara, S., Reinhart, G., Sauer, O., Schuh, G., Sihn, W. and Ueda, K. (2016). Cyber-physical systems in manufacturing, *CIRP Annals*, 65, 621–641.

Niebel, B.W. (1965). Mechanized process selection for planning new designs. *ASME Paper*, 737.

Phanden, R.K., Jain, A. and Verma, R. (2011). Integration of process planning and scheduling: A state-of-the-art review. *International Journal of Computer Integrated Manufacturing*, 24(6), 517–534.

Phanden, R.K., Jain, A. and Verma, R. (2012). A genetic algorithm-based approach for job shop scheduling. *Journal of Manufacturing Technology Management*, 23(7), 937–946.

Phanden, R.K., Saharan, L.K. and Erkoyuncu, J.A. (2018). Simulation Based Cuckoo Search Optimization Algorithm for Flexible Job Shop Scheduling Problem. *In the Proceedings of 2018 International Conference on Intelligent Science and Technology, ICIST 2018*, London, UK, June 30–July 2, pp. 50–55.

Pinedo, M. (2008). *Scheduling: Theory, Algorithms and Systems*, Prentice-Hall, Upper Saddle River, NJ.

Saygin, C. and Kilic, S.E. (1999). Integrating flexible process plans with scheduling in flexible manufacturing systems. *International Journal of Advance Manufacturing Technology*, 15, 268–280.

Sugimura, N., Hino, R. and Moriwaki, T. (2001). Integrated Process Planning and Scheduling in Holonic Manufacturing Systems. *In the Proceedings of IEEE International Symposium on Assembly Task Planning, Soft Research Park*, Fukuoka, Japan, 4, pp. 250–254.

Tulkoff, J. (1987). Process planning in the computer-integrated factory, *CIM Review*, 4(1), 24–27.

Wang, L. and Zheng, D.Z. (2002). A modified genetic algorithm for job shop scheduling. *International Journal of Advance Manufacturing Technology*, 20, 72–76.

Wolfe, P.M. (1985). CIMS series. 20. Computer-aided process planning is link between CAD and CAM. *Industrial Engineering*, 17(8), 72.

Zhang, H. and Alting, L. (1994). *Computerized Manufacturing Process Planning Systems*, Chapman & Hall, London.

Zhang, S. and Wong, T.N. (2016). Studying the impact of sequence-dependent set-up times in integrated process planning and scheduling with E-ACO heuristic. *International Journal of Production Research*, 54(16), 4815–4838.

2

Approaches to Integrate Process Planning and Scheduling

Rakesh Kumar Phanden

Amity University Uttar Pradesh

Ajai Jain

National Institute of Technology Kurukshetra

CONTENTS

2.1 Introduction 19
2.2 Classification of IPPS Approaches 20
    2.2.1 NLA 20
    2.2.2 CLA 28
    2.2.3 DA 31
    2.2.4 Other Approaches 38
2.3 Analysis of Literature Review 41
2.4 Conclusion 42
References 43

2.1 Introduction

The manufacturing scenario is drastically changing day by day in order to achieve various objectives such as reduction of manufacturing lead time, production of a vast variety of products, meeting standards of product quality and customers' needs at a viable cost, and customised (make-to-order) demands of customers, as well as to reduce energy consumption and pollutant emissions (Zhang et al. 2016; Gindy et al. 1999). It is difficult to achieve these objectives in the traditional manufacturing systems because of various problems in the existing segregated functioning of production scheduling and process planning, such as; (i) the process planner assumes that the shop floor resources are available with unlimited capacity at all times and he generates the process plan considering the utmost recommended resource repeatedly; (ii) the constraints considered during process planning may vary during scheduling because of the time delay in both the activities (Baykasoğlu and Özbakır 2009; Sugimura et al. 2001; Usher and Fernandes 1996). Consequently, it generates an unrealistic, unoptimised (or suboptimised), and even invalid, process plan which cannot be freely applied on production. These problems of a traditional manufacturing system can be solved by *Integration of Process Planning and Scheduling* (IPPS). The integrated manufacturing approach performs well in the present-day manufacturing scenario that facilitates manufacturing flexibility (optimum utilisation of available alternative resources) and generates a practical process plan that can be implemented on shop floor, readily (Liu et al. 2018; Phanden et al. 2011; Guo et al. 2009; Kumar and Rajotia 2002, 2003; Lee and Kim 2001; Gindy et al. 1999; Saygin and Kilic 1999). Hence, the IPPS is important to attain the ultimate integrated manufacturing system and to eradicate the traditional manufacturing approach. IPPS can be achieved in various ways like *non-linear approach* (NLA), *closed-loop approach* (CLA), and *distributed approaches* (DAs) proposed by various researchers across the world. In the next section, the different IPPS approaches available in the literature have been discussed in depth.

2.2 Classification of IPPS Approaches

The straight way to solve the IPPS problem is by combining the process planning and scheduling functions into one problem, although the process planning as well as scheduling functions alone fall into the category of *nonpolynomial* (NP) - hard problems. Also, these are combinatorial optimisation problems. Thus, the idea of combining both functions into one leads to an additional NP-hard problem (Baykasoğlu and Özbakır 2009). This approach also needs complete dismantling and reorganisation of the existing facilities in the company. Hence, merging of process planning and scheduling functions is not a viable method and it is extremely difficult to apply this new method to an existing set of process planning and scheduling departments of a company. In this direction, Tan (1998) attempted to solve the IPPS problem using a mathematical programming approach with limited success (Tan and Khoshnevis 2000). On the other hand, IPPS can be achieved by exchanging the information amid the process planning and scheduling functions. Based on the idea of information exchange, various categories have been proposed in the literature (Baykasoğlu and Özbakır 2009). However, the three main IPPS approaches are NLA, CLA, and DA. These approaches and their corresponding contributions are presented below.

2.2.1 NLA

NLA is also known as *multi-process planning* or *flexible process planning* or *alternative process planning* or *non-linear process planning* approach. In NLA, *Multiple Process Plans* (MPP) are created for each job by considering the operational, sequencing, and processing flexibility in manufacturing. Operational flexibility means the prospect of processing an operation of a job on multiple machines available on shop floor, and sequencing flexibility means the prospect of exchanging the required sequence of manufacturing operations with an alternative sequence of operations. Processing flexibility includes the prospect of manufacturing an identical feature of the product by an alternative set of operations or via a sequence of operations. MPP improves the manufacturing flexibility of the system. NLA is the most basic and simplest approach for IPPS (Jain et al. 2013; Baykasoğlu and Özbakır 2009; Tonshoff et al. 1989; Gindy et al. 1999). The idea behind NLA approach is that all problems that could be solved before the commencement of the manufacturing stage should be solved beforehand. Therefore, NLA approach considers the static conditions of the shop floor (Baykasoğlu and Özbakır 2009; Zhang and Merchant 1993). The process planning database consists of all the possible process plans of a production order. In this database, the MPP of each job is arranged in a specific hierarchy as per the process planning criteria like total minimum production time, setup time, transportation time, etc. Hence, the process plan ranked at the top position is prioritised to submit for scheduling. In case the first process plan is not acceptable as per the existing factory floor conditions, then the next process plan will be supplied in order to perform scheduling. Subsequently, this iterative procedure will be executed until a workable process plan is selected from the process planning database. Thus, the scheduling function makes the real decision of selection of the process plan. Moreover, the scheduling function selects the process plan based on due dates of the production order, batch size, and other optimisation criteria such as manufacturing lead time, throughput, makespan, etc. Figure 2.1 presents the flowchart of the NLA approach.

The NLA approach is restricted to only one-direction of information flow, from shop floor to process planning and then to scheduling. Thus, it is a one-time update of shop floor status (after completing a production order); consequently, it is difficult to obtain a fully optimal IPPS solution. However, an advanced manufacturing system is capable to maintain and update the MPP continually. Therefore, NLA seems suitable for IPPS (Baykasoğlu and

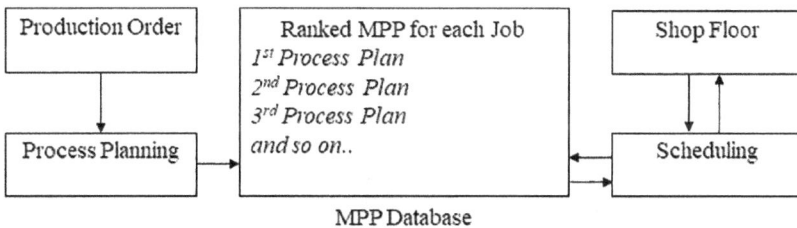

FIGURE 2.1
Flowchart of NLA approach (Phanden et al. 2011; Li et al. 2010b; Zhang and Merchant 1993).

Özbakır 2009; Kim and Egbelu 1999; Kempenaers et al. 1996). Moreover, it can easily be applied in a factory setting without disturbing the existing facilities of process planning and scheduling departments. Although, when the number of products is increased, the number of process plans (MPP) also increases quickly and by a large amount, which leads to cache intricacy. Moreover, some process plans from the MPP database may not be valid as per the existing shop status, thus the consideration of all the possible manufacturing flexibilities for resources assignment may raise the intricacy of process plans representation (Huang et al. 1995; Zhang and Merchant 1993).

The idea of IPPS originated from Chryssolouris and Chan (1985). They introduced *Manufacturing Decision Making Approach* (MADEMA), which considered alternative resources during production. The decision matrix was designed to represent available alternative resources for operation. The row vectors showed alternative resources while the columns contained their attributes, and the cell entries were filled with values of attributes for the corresponding alternative resource. It contained the following steps: (a) *determine alternatives*, (b) *determine attributes*, (c) *determine consequences with respect to attributes for each alternative*, (d) *apply decision rules for choosing the best alternative* and, (e) *select the best alternative*. The decision of selecting the best alternatives was made as per the linear attributes grouping through weights, and/or alternative resources having more possibilities in order to yield a better value of utility index.

Sundaram and Fu (1988) proposed a production scheduling procedure for a job shop based on the output of process planning to minimise makespan and to equilibrize machine loading levels. The authors adopted a *Group Technology* (GT)-based automated system to improve schedule. The system was named as *Integrated Computer-Aided Process Planning and Scheduling* (ICAPPS). The authors developed a GT-based scheduling heuristic, termed *Key Machine Loading*. It was linked to two process planning algorithms, namely *Process Planning Generator* and *Operation Planner*. The idea behind this approach was to keep a machine loaded with jobs so as to keep it busy continuously.

Tonshoff et al. (1989) presented FLEXPLAN to perform IPPS, which creates MPP before the commencement of production. The best process plans were selected as per the availability of machines in the scheduling phase. Also, they included responsive re-planning to accommodate disturbances during production.

Jablonski et al. (1990) developed an IPPS system called *Flexibly Integrated Production Planning and Scheduling* (FIPS) with trio segments. Foremost was *Feature Recognition, Extraction, Decomposition, and Organisation System* (FREDOS), next was *Static System Manager* (SSM), and last was *Dynamic Resource Scheduler* (DRS) to identify geometric features, to generate all possible process plans combinations, and to select the best combination as per user-defined strategy, respectively. They concluded that reactive scheduling is feasible with feature-based process planning.

Concurrent Manufacturing Planning and Shop Control for Batch Production (COMPLAN) was a European ESPRIT project 6805 from 1992 to 1995. The COMPLAN project involved the development of a software which carried out automatic as well as manual process planning, and scheduling activities to manufacture small-size batch orders in a job shop environment. They achieved IPPS using MPP. It owned *Process Planning Module* (PPM) and *Workshop Scheduling System* (WSS) to create MPP based on anticipated resource load, and designated available alternative resources while scheduling. This project was extended from FLEXPLAN, proposed by Kruth and Detand (1992).

Palmer (1996) solved NLA-based IPPS using *Simulated Annealing* (SA) algorithm. They altered the configurations of operations in three ways, viz., (i) by inverting two consecutively ordered operations on machines, (ii) by inverting two consecutively ordered operations for the jobs and, (iii) by changing the technique applied to do the operation. They quantified the system effectiveness with respect to the tardiness and mean flow time, as well as makespan measurements. Also, they compared the SA algorithm with dispatching rules and found that the former was outperformed.

Kim and Egbelu (1999) presented a mathematical model for production scheduling using MPP in a job shop to minimise makespan and throughput. *Process Plan Selection Subsystem* (PPSS) and *Shop Scheduling Subsystem* (SSS) were designed to select process plans and to generate viable schedules, respectively. Both subsystems worked in an iterative manner to trade-off amid the required schedule and the possibility of modification in existing process plan. They developed two scheduling algorithms called *pre-processing* and *heuristic/iterative*, based on collective advantageous characteristics of a branch and bound method, integer programming as well as *Earliest Completion Time* (ECT) rule. Authors have drawn important conclusions from computational time viewpoint such as, (i) the run-time via *pre-processing* method was less than *mixed integer programming* (MLP) method but more than *heuristic model*; (ii) when quantity of parts raise the quality of solution attained through algorithms was poor, but when the number of machines increases there was no clear effect on the solution quality attained by heuristics; (iii) when quantity of process plans for each job was increased there was an adverse impact on the performance of algorithms.

Aldakhilallah and Ramesh (1999) presented *Computer-Integrated Process Planning and Scheduling* (CIPPS) framework having *Super Relation Graph* (SRG), *Cover Set Model* (CSM), and *Cover Set Planning and Scheduling* (CSPS) modules for automated feature recognition, determination of minimal cover sets, and determination of feasible process plan as well as feasible cyclic schedule, respectively. This architecture was composed of three operational modes, namely, *Dynamic Support for Design Decision* (DSDD), runtime *Intelligent Operational Control* (IOC), and *Data Consolidation and Integration* (DCI). The DCI acts as an interfacing mode and integrates CIPPS with other manufacturing functions. IOC mode updates the changes occurred on the shop floor and DSDD mode helps decision-making during the design phase.

Weintraub et al. (1999) presented a scheduling procedure using alternative process plans to meet the due date within minimum manufacturing cost. They developed a virtual factory (a Windows-based software) that possessed a simulation-based scheduling algorithm to minimise the lateness of jobs. Further, they applied *Tabu Search* (TS) to reduce lateness through process plans with alternative routings. They selected only two alternative process plans per job, based on current shop floor status. They drew a noteworthy conclusion that in case of varying status of shop floor, the availability of alternative process plans plays an important role in meeting the due dates of the delivery of products.

Saygin and Kilic (1999) framed an integration approach considering MPP and offline (predictive) scheduling, using mathematical and heuristic-based algorithms to minimise completion time in *Flexible Manufacturing System* (FMS). It possessed four modules, viz., (i) *machine tool selection module*, (ii) *process plan selection module*, (iii) *scheduling module*, and (iv) *re-scheduling module*. *Dissimilarity Maximisation Method* (DM) was applied to select suitable process plans for the part mix. The authors stated that the optimised process plan based on *Shortest Processing Time* (SPT) dispatching rule and/or a minimum number of operations may not ensure a good outcome from the system.

Lee and Kim (2001) presented a simulation-based *Genetic Algorithm* (GA) for IPPS. GA provided a process plan combination that was simulated using SPT and *Earliest Due Date* (EDD) rules to measure makespan and lateness performance measures. Authors achieved a drop of 20% in makespan as compared to randomly selected process plan combinations for a given production order.

Yang et al. (2001) presented a scheduling certified model as per the part features and alternative process plans, which possessed four components, viz., (i) *relational manufacturing database*, (ii) *form feature recognition*, (iii) *process alternative generation*, and (iv) *scheduling state evaluation*. Graph-based, as well as rule-based algorithms, were used to extract the manufacturing features of part design and to generate MPP. Every process plan was tested on the scheduling system and a key process plan was selected based on the approaching due date. The authors stated that the system is capable to choose a process sequence according to delivery time commitments of the production order.

Moon et al. (2002) developed a GA-based IPPS system to fit into a manufacturing environment containing multiple plants in the supply chain to minimise total tardiness. They used the *Travelling Salesman Problem* (TSP) formulation to find operation sequencing for jobs. They compared the proposed approach with TS and found that GA-based approach was more effective in terms of computation time consumption. In addition, they found that the number of generation and size of population are two key aspects that affect the optimised output of developed GA-based IPPS system

Kim et al. (2003) proposed *Symbiotic Evolutionary Algorithm* (SEA) in order to achieve IPPS in FMS. The authors stated that the idea behind the proposed approach was to search for various pieces of the solution in parallel rather than searching the entire solution in a single search. The proposed approach effectively embraced processing flexibility, operational flexibility as well as sequencing flexibility while optimising makespan and mean flow time performance measures.

Zhao et al. (2004) presented GA for IPPS in job shop manufacturing in order to minimise processing cost, number of parts rejection, and makespan. They picked alternative machines using *Fuzzy Interface System* (FIS) from fuzzy logic toolbox of MATLAB®. GA and GT algorithms (Giffler and Thompson 1960) were applied in order to stabilise the machine loads and to evaluate fitness values of chromosomes, respectively. They extended this work by applying the *Particle Swarm Optimisation* (PSO) method to achieve balanced loading on machines effectively (Zhao et al. 2006). In addition, they applied the proposed IPPS system in *Holonic Manufacturing System* (HMS) and used *Differential Evolution* (DE) algorithm as well as hybrid PSO for balancing the load on machines precisely (Zhao et al. 2010). Choi and Park (2006) presented GA for IPPS in order to optimise the makespan fitness function for a job shop manufacturing environment with alternative operational sequences as well as substituting machines flexibility.

Jain et al. (2006) presented an IPPS model to avail the benefits of manufacturing flexibility available at factory floor level by generating the MPP for each job. The proposed model was designed to sync on existing facilities of the planning and scheduling divisions of a company. The proposed scheme possessed two modules, namely, *Process Plans Selection Module* (PPSM) and *Scheduling Module* (SM) in order to select the top four process plans for an individual job, as well as to generate the schedule by opting for the best process plan, respectively. They concluded that the availability of MPP was helpful to minimise mean flow time as well as makespan performance measures during FMS scheduling.

Li and McMahon (2007) proposed SA-based approach in order to optimise IPPS problem for a job shop configuration for makespan, tardiness, and to balance machine utilisation levels as well as to reduce manufacturing cost. The explored the search space by considering operations, processing and sequencing manufacturing flexibilities available at shop floor. They designed two groups of data structures, viz., to show process plans, and to lay down a schedule of the part mix. They compared the performance of the developed algorithm with PSO, GA, as well as TS algorithms, and found that it performed satisfactorily. Also, the proposed model was adjustable to opt for different performance objectives in order to meet everyday needs.

Wang et al. (2008a) proposed an IPPS method that possessed reactive scheduling capability (for machine breakdown) in batch manufacturing. They achieved integration over revising process plans in search domain in an iterative manner. This approach contained heuristics to minimise tardiness as well as to hold the cost of manufacturing. They used SA algorithm to optimise process plan for prismatic parts. The authors concluded that the cost of manufacturing (after process plan modification trade-off) was maintained at a low level but tardiness improved.

Moon et al. (2008) presented the IPPS approach in supply chain using Topological Short Technique to generate a set of viable process sequencing to optimise makespan. MIP model was framed that considered alternative resources, sequences, and precedence constraints. Through the results of their experiments, the authors claimed that the presented IPPS model was proficient in order to crack complex and sizable problems in supply chain involving IPPS.

Li et al. (2008b) proposed GA in order to optimise IPPS problem in a job shop for makespan performance measure. They designed effective genetic representation and GA operators. The chromosome was divided into two strings, viz., process plan string and scheduling plan string. The authors concluded that makespan performance of the system was inferior while deprived of integration, as compared to the presented IPPS methodology. Li et al. (2010c) synthesised GA with TS in order to solve IPPS for makespan performance measure. They designed three strings chromosomes to put up alternative process plans, scheduling plans, and corresponding machines number. Moreover, the third-string was designed to select the machine sets corresponding to operations. The authors stated that this approach was capable to find an optimal solution for IPPS. Moreover, Li et al. (2012a) presented *Game Theory*–based blended techniques (using GA and TS algorithm) with Nash equilibrium to solve multi-objective IPPS. Li et al. (2012b) presented *Active Learning Genetic Algorithm* (ALGA) in order to solve IPPS problem. ALGA possessed better searching ability than regular GA. They designed a crossover operator as a learning operator that learned useful data from a healthier solution (chromosome) only. They stated that the presented IPPS approach was promising and intelligent enough to optimise makespan performance measure.

Haddadzade et al. (2009) presented IPPS for prismatic parts in a job shop. It consisted of PPM and SM, to generate MPP and to generate a schedule with respect to manufacturing cost as well as due date, respectively. The authors stated that the presented IPPS method was capable to meet different objectives like due dates and minimum cost-dependent overtime.

Baykasoğlu and Özbakır (2009) presented the *Multiple Objective Tabu Search* (MOTS) algorithm for IPPS to minimise the process plan cost as well as the flow time. They proposed a *Generic Process Plan* (GPP) generator and a dispatching rule-based heuristic in order to produce a process plan and a feasible production schedule, respectively. The authors concluded that the manufacturing flexibility and process plan cost were varying and inversely proportional.

Leung et al. (2010) solved the IPPS problem using *Multi-Agents System* (MAS) approach for makespan performance measure. They framed *Ant Colony Optimisation* (ACO) algorithm over MAS platform for job shop manufacturing system. The proposed algorithm considered processing flexibility through AND/OR graphs and extracted possible MPP. They concluded that MAS-based ACO approach is promising for IPPS.

Naseri and Afshari (2012) proposed hybrid GA for IPPS with precedence constraints on the alternative sequence of operation to minimise makespan performance. They used graphs in order to illustrate devious priority connections between operations to generate MPP. Initially, they formulated the mathematical model for *Flexible Job Shop Scheduling Problem with Alternative Process Plans* (FJSP-APP) using *Mixed Integer Linear Programming* (MILP). The authors claimed that the proposed approach was capable of generating a viable process route for each part considering non-linear precedence relation and to assign operations to suitable machines, as well as sequencing the operation on each machine, simultaneously.

Haddadzade et al. (2014) presented IPPS for job shop with stochastic processing time. The proposed system generated MPP and four optimal process plans were selected using Fijkstra algorithm. They used a hybrid algorithm composed of SA and TS to solve the formulated mathematical model within a reasonable time. They tested the proposed algorithm in a deterministic and stochastic way. The comparison shows that the results with the deterministic environment were weak as compared to the results with stochastics data.

Wang et al. (2014) proposed AND/OR graph-based ACO algorithm for IPPS to minimise makespan. They presented IPPS on AND/OR graph in which the nodes showed the operation of jobs and the arcs showed a possible route between the nodes. The authors stated that the idea behind the proposed approach was that the ant colony travels through essential junctions on the graph from the initial point to the finishing point in order to find a suboptimal result.

Jin et al. (2016) proposed *Multi-Objective Memetic Algorithm* (MOMA) for IPPS to optimise makespan and workload on machines. They combined *variable neighbourhood search* (VNS) and an *objective-specific intensification search method* for each given objective to improve all possible schedules. Also, TOPSIS procedure was applied to choose a reasonable schedule from an optimal Pareto front. The authors concluded that MOMA was effective to improve the given multi-objectives and it outperformed the NSGA-II algorithm for multi-objective IPPS problem.

Xia et al. (2016) presented GA & VNS-based hybrid approach for dynamic IPPS model that considered fresh production orders as well as machine breakdown disturbances in the system. They applied rolling scheduling scheme and two neighbourhood schemes to explore local search. The authors concluded that the proposed approach outperforms GA.

Petrović et al. (2016a) proposed nature-inspired *Ant Lion Optimisation* (ALO) algorithm that works based on the intelligent behaviour of antlions to hunt ants, to solve the IPPS problem. They considered machine flexibility, sequence flexibility, and process flexibility through AND/OR network. The optimal sequence of operation was determined based on minimum production time criteria and the optimal schedule was generated based on minimum makespan performance measure. Authors concluded that ALO was a promising approach to solve the IPPS problem. Petrović et al. (2016b) presented PSO and chaos theory-based hybrid algorithm for IPPS and compared it with many benchmark approaches.

Zhang et al. (2016) proposed NLA-based IPPS approach using GA that minimises the energy consumption of machining tools in the manufacturing system using Therblig-based model. They stated that the proposed approach was effective enough to save energy consumption at the shop floor, resulting in green low-carbon discharge.

Sobeyko and Mönch (2017) proposed MIP formulation for IPPS in a large-scale flexible job shop to optimise weighted tardiness. The VNS method was used at the process planning level with appropriately designed neighbourhood structures and heuristics were used at the scheduling level. The authors considered that varying bills of materials and routes can be used to manufacture a product in different ways. They concluded that the proposed approach proved better than GA and memetic algorithms to solve large problems rapidly. Shokouhi (2018) proposed a GA-based IPPS model for flexible job shop in view of precedence relationship among operations, that optimise weighted-sum multiple objectives (makespan, critical, and total machine workload) in Pareto space.

2.2.2 CLA

CLA is a dynamic and real-time feedback mechanism-based IPPS approach. Here, the process plans are generated after considering the availability of resources and scheduling objectives. Both, process planning and scheduling systems, can interact to generate a viable process plan based on the real-time status of production facilities. In other words, the scheduling system of CLA approach is capable to access the existing status (availability) of machines on the shop floor in order to assign it to the incoming job to achieve the desired scheduling objectives such as makespan, delivery due date, tardiness, etc. In this approach, every process plan is feasible to execute on the shop floor readily, without alteration. Every time an operation is completed on the shop floor, a feature-based workpiece description is studied to determine the next operation and allocate the resources. Thus, the dynamic updates from factory level play a crucial role to create a real-time-based process plan in CLA. It can also be termed online IPPS approach. Figure 2.2 presents the flowchart of the CLA approach.

CLA is recognised as a real IPPS integration approach. However, the implementation of CLA-based IPPS needs complete reorganisation and dismantling of the existing process planning and scheduling departments of the company. In addition, it needs powerfully configured software and hardware devices. It involves a huge number of reworking steps at the local view to schedule each job and machines after completing each operation, this iterative procedure confines the solution space for succeeding operations

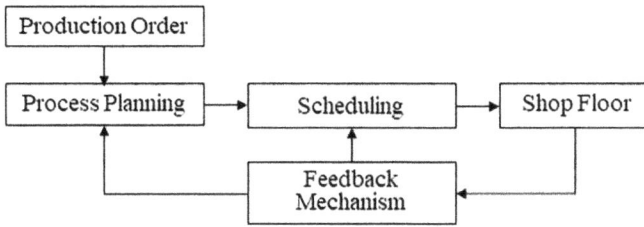

FIGURE 2.2
Flowchart of CLA approach (Zhang and Merchant 1993; Li et al. 2010b; Phanden et al. 2011).

(Zhang and Merchant 1993; Gindy et al. 1999; Baykasoğlu and Özbakır 2009). However, the available manufacturing flexibility cannot be overlooked to generate real-time-based viable process plans.

Khoshnevis and Chen (1991) presented an idea of dynamic feedback mechanism for IPPS. They addressed the issue of IPPS and demonstrated a methodology, while also presenting an example to accomplish it. Dong et al. (1992) presented a dynamic IPPS approach that extracts product features by considering shop floor status and creates a *rough process plan* (RPP). They explored all the manufacturing methods available for each product feature which can process in one machine setup. In RPP, the priority of manufacturing for each operation was decided based on geometric constraints. RPP with available alternatives was a regular input to perform scheduling. Moreover, the *Smallest Slack Time* (SST) dispatching rule was used to schedule a batch manufacturing in a shop.

Usher and Fernandes (1996) presented *Process planning ARchitecture for Integration with Scheduling* (PARIS), a feature-based planning system that worked in dual planning sections, namely, (i) *static planning* and (ii) *dynamic planning*. The former section selects and assigns the sequences of operations to the machines existing in a shop and in the later phase the machines were considered based on their availability and the desired scheduling objectives. The authors stated that the real-time planning phase of PARIS introduces a drop in work-load during the second phase, and it was capable to process both rotational and prismatic components.

Cho et al. (1998) presented *Block Assembly Process Planning and Scheduling* (BAPPS) prototype for ship-building. It was composed of different modules, namely, *block assembly PPM, SM, bottleneck block selection module*, and *process re-planning module* that worked on the initial process plan in a closed and iterative manner to balance workloads on machines. They applied rule-based reasoning to generate the initial process plan and to find optimal assembly sequences and assembly units. The rescheduling algorithm works to redistribute blocks over the substitute assembly shops and the bottleneck periods were found by a heuristic based on the entropy-based parturition scheme.

Kumar and Rajotia (2003) presented a scheduling method with *Computer-Aided Process Planning* (CAPP), to generate online process plan considering the availability of machines and substitute process routing in a job shop system to optimise the tardy jobs and flow time together. They designed a scheduling feature to assign the operations to machines. They found that the developed method efficiently assigned the loads on machines in an online network. In Kumar and Rajotia (2006), the capacity and cost of machines were provided for the assignment of operations over the machines. It was occupied with dual modules, namely, a schedule plan generating module, and a process plan generating module. These modules were networked side-ways *Decision Support System* (DSS). In addition, DSS accessed tools and work materials, machine tools, and machining parameters database files. The scheduling factor proposed by Kumar and Rajotia (2003), was adopted for the allocation of setups to machines.

Zhang et al. (2003a) presented the IPPS model to perform batch manu-facturing of prismatic parts using GA and SA algorithms. In this system, the PPM explored entire search domain build on the operation for a job to determine the optimal process plan. The integration was attained when the *Intelligent Facilitator* module was able to modify the process plans after receiving feedback from PPM for a selected job. Authors presented *machine-utilisation* and *tardy-job* heuristic algorithms to perform scheduling. This approach was extended by Wang et al. (2008b). They developed *Fine-Tuning* and *Quick-Tuning* heuristics to explore the solution domain for designated operations of a chosen tardy job through an iterative manner. They stated that the presented algorithms were good enough to reduce job tardiness.

Sugimura et al. (2001) presented an IPPS for HMS to optimise total machining time and setup time using *Dynamic Programming* (DP) and GA. They performed real-time scheduling in which the machine schedule was generated by selecting a suitable machine for an individual job to carry out subsequent operations dynamically. In Sugimura et al. (2003), the proposed approach was extended and considered minimisation of machining cost and shop time. Sugimura et al. (2007) presented an IPPS model for HMS. Here, process plans were modified through DP and GA to reduce makes-pan, tardiness, and machining cost. An optimised combination of process plans was selected using GA. SPT, *Apparent Tardiness Cost* (ATC), and *Total Work Remaining* (TWR) dispatching rules were used to schedule jobs. They designed scheduling holons and job holons for scheduling and process plan-ning, respectively. Initially, the job holons generate optimised process plans and send a set of process plans to the scheduling holon. Subsequently, sched-uling holon generates schedule for the combination of process plans received from the job holon. Afterwards, the scheduling holon passes the response to each job holon. Subsequently, the job holons update their process plans owing to the response received. Thus, the iterative procedure is followed until an optimised combination of schedule and process plans is attained.

Wong et al. (2006a) proposed MAS-based IPPS in a job shop environment, that was composed of the machine and part agents. These agents accessed operational details from the AND/OR graphs to negotiate and generate a schedule. They developed a currency conversion function which includes due dates and processing times for the bidding process. The authors used Java-based simulation architecture called *Multi-Agents Negotiation* (MAN) to execute the proposed IPPS scheme. They drew an important conclusion that the proposed approach performed better than meta-heuristics from a flow time performance measure viewpoint. In addition, they extended the proposed scheme by adding a supervisor agent (Wong et al. 2006b). They called it *Online Hybrid Agents-based Negotiation* (oHAN). The supervisor agent acted as a *system coordinator manager* for global re-scheduling objective and influenced the decisions made by local agents (machine agent and part agent). The authors stated that the developed model was useful for solving large size re-scheduling problems and it also yielded an excellent overall solution. The authors suggested an extension in the proposed scheme by introducing a mobile agent to deal with the production order details.

2.2.3 DA

DA is also known as *concurrent process planning* or *collaborative process planning* or *just-in-time process planning* or *progressive process planning* or *phased process planning* approach. As the name represents, DA works in two dispersed and parallel phases of process planning and scheduling. The first phase is called *preplanning phase,* and the second is called *final planning phase.* In the former phase, the product design data are analysed to extract the required process planning information. Also, the feature relationship and corresponding manufacturing operations are addressed. In addition, the functionality of the required machine and available manufacturing flexibility, as well as the relationship between the different jobs is also analysed. In the latter phase, the process plan is edited as per the existing status of shop floor. The job operation is matched with the operational capability of the available machine. The point of integration arises when the machine is available, and the job is ready to process. This procedure results in real-time event management to seize the constraints of process planning and scheduling. Figure 2.3 presents the flowchart of the DA approach.

DA has been recognised as the only approach which is capable of integrating technical and capacity-related planning task into a dynamically fabricated IPPS system. Although, it needs a powerfully configured hardware and skilled software application from an implementation viewpoint. Also, the space of DA approach is restricted with certain CAPP functions like machine and process selections, because the detailed process planning task has been pushed down towards the manufacturing stage to improve flexibility. In addition, the existing process planning and scheduling departments of a traditional manufacturing company need to reorganise and dismantle to implement the DA-based IPPS system (Zhang and Merchant 1993; Gindy et al. 1999; Morad and Zalzala 1999; Baykasoğlu and Özbakır 2009; Li et al. 2010b; Phanden et al. 2011).

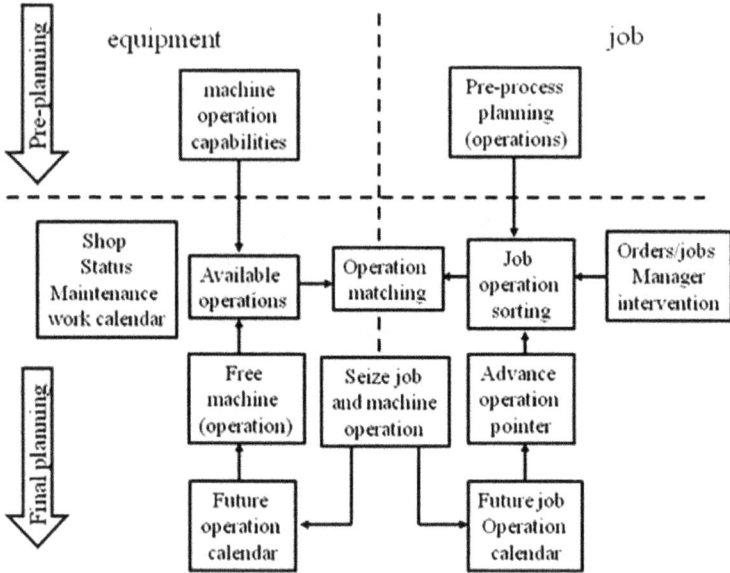

FIGURE 2.3
Flowchart of DA approach (Phanden et al. 2011; Zhang and Merchant 1993).

Aanen et al. (1989) presented a systematically organised IPPS scheme for FMS to satisfy the due date and to minimise the idle time as well as the change-over time of resources. They developed *Supervisory Control System* (SCS) software to integrate various components of FSM. They first solved process planning, followed by scheduling. If the generated schedule was not acceptable then the information was fed back to the planning level iteratively. They have scheduled machining activities as well as operator activities for a period of ten days at the planning level and splits for a day (called day list). Each day, the machining activities were scheduled and makespan value was optimised for a day list utilising *Branch-and-Bound* algorithm.

Zhang and Merchant (1993) extended the idea of alternative process plans and dynamic feedback and proposed the method of hierarchical IPPS. They presented *Integrated Process Planning Model* composed of: (i) *pre-planning module*, (ii) *pairing planning module* and, (iii) *final-planning module* placed at initial integration, decision-making, and functional integration levels, respectively. The first module performed the features of logic building, setup diagnosis, as well as process identification. The output was in the form of machining operations and time accompanied and most feasible setups recommendations. The second module selected machines, tooling, fixtures and exact processing time associated. The third module performed a study on tolerance of operations, sequencing of operations as well as total cost and time investigations. The decision-making stage matches the equipment (resources) with processes and setups available. This module acts as a prime element that performs based on the real-time data.

Huang et al. (1995) presented a progressive IPPS approach which consisted of pre-planning phase (after product design), pairing planning phase (after release of the production order to shop floor), and final planning phase (before start of manufacturing), working in different time horizons. It was an all-phases integration model that contained PPM and SM for process plan generation according to minimum manufacturing lead time, and allocation of resources, respectively. They stated that the presented approach was capable to deal effectively with computation complexity. In addition, they recommended that elements such as machine cell monitoring and database management, as well as the knowledge processing layout, must be in the script to implement the proposed approach.

Kempenaers et al. (1996) presented collaborative IPPS that composed of *Process Plan Evaluation* (PPE), *Schedule Evaluator* (SE), and *Workshop Evaluator* (WE), to enhance the quality of delivery output and setup of feedback information mechanism at all evaluation levels using an *Evaluation Module* (EM). WE evaluates the various information like operation time, machine breakdown, process plans, and other disruptions from shop floor, to take appropriate actions. SE compares the performance of generated schedule with the output result of shop floor. Thus, an alternative resource is added in the selected process plan to replace broken and/or unavailable resource. Hence, the production constraints projected by SE were considered by PPE.

Mamalis et al. (1996) introduced an online IPPS system that composed of *offline process planning and scheduling phase* and *online process planning and scheduling phase*. In offline phase, MPP was generated by the bi-directional flow of information (dynamic exchange of information) between CAPP and scheduling system. In online (decision-making) phase, disruptions of shop floor were considered using *Decision Making Module* (DMM). DMM modified the initial process plan and generated an optimal schedule based on the SPT, *First-Come-First-Serve* (FCFS), and *Longest Processing Time* (LPT) to minimise the total delay of jobs. This system was authenticated through simulating the job shop machining environment for rotational parts only. The authors stated that the proposed online IPPS system is feasible to consider the dynamic behaviour of job shop.

Gu et al. (1997) presented the bidding-based MAS for IPPS, which was composed of *part agent*, *shop manager agent*, *machine agent*, and *tool agent*. The machine agent performed STEP parsing and interpretations to generate process planning data, operations and setup planning, tolerance analysis, and all other functions related to process planning, while the other agents worked on the normal activities with respect to the task and resources to which they are characterised. The decisions were made through negotiation among setup and machining time, tool changing time, the due date of order, as well as the cost of tooling. The authors stated that the proposed bidding-based MAS was practical to achieve IPPS.

Sadeh et al. (1998) presented *Integrated Process Planning/Production Scheduling Shell* (IP3S) using blackboard structure based on the concurrent and progressive modifications in process plans for IPPS in an agile manufacturing environment. IP3S was composed of a controller, a blackboard, a collection of knowledge sources, and what-if based GUI. They designed different knowledge sources such as knowledge sources for process planning, production scheduling, communication, and analysis, to perform the associated functions. The proposed architecture performed cyclically to resolve issues using a suitable information database.

Gindy et al. (1999) presented concurrent engineering-based IPPS method that composed of feature-based process planning, knowledge-based facility modelling, and simulation-based scheduling module. They proposed a *Resource Element* (RE) concept to characterise the manufacturing system. Form generating schema was used to describe process capability. Alternative features were extracted during process planning. A set of REs described the machines of a manufacturing system. A GPP (machine-independent) was an input to scheduling system to minimise tardiness and to balance machine utilisation levels. Just before the commencement of manufacturing, the final process plan and production schedule were generated in parallel. Authors compared the performance of RE-based and machine-based traditional systems for mean flow time and tardiness. They drew various interesting facts such as (i) RE-based system was less sensitive towards the due date assignment and dispatching rule approaches, (ii) RE-based system was sensitive to disruption due to machine breakdown and frequent changes in product demand, (iii) the average machine utilisation for machine-based system was 22% and the same for the RE system was 40%, (iv) *Total WorK content* (TWK) scheme was proved to be a superior approach from the mean tardiness performance measure viewpoint of scheduling.

Morad and Zalzala (1999) presented concurrent engineering-based GA for IPPS that simultaneously optimised the processing capabilities of machines (with respect to quantity of rejections and manufacturing cost associated with alternative machines) and scheduling of jobs using SPT dispatching rule. They designed chromosome with sequence of operations, jobs, and machines to determine fitness for multiple objectives of minimising makespan, cost of production, and quantity of rejections produced. They concluded that the proposed multi-objective weighted-sum approach performed better than a traditional *multi-objective genetic algorithm* (MOGA) for makespan and quantity of rejections produced.

Chan et al. (2001) proposed *Integrated, Distributed and Co-operative Process Planning System* (IDCPPS) based on the MAS that composed of initial planning level (first), decision-making level (second), and detailed planning level (third). At the first level, a set of alternative process plans were generated and sent to the second level. At the second level, the current shop floor status was considered for scheduling and a set of ranked-cum-optimal alternative process plans were generated. At the third level, linear process plans were generated that considered a detailed selection of tools, machines, parameters, and calculations for machining time and cost. The proposed scheme integrates both functions at each level. They concluded that IDCPP was a CAD tool to participate in down-stream constraints during designing and it integrates process planning and scheduling. It also responded dynamically to changes that happened at the shop floor with the help of MAS, in the later phases.

Wu et al. (2002) presented a Java-based MAS architecture in a virtual production environment for IPPS. They designed a cost function to select an optimal partner (resource) in virtual manufacturing based on processing time consumption, locations, and manufacturing capability of partners, as well as due date of jobs. The proposed system composed of an enterprise phase and a partner phase to select optimal manufacturing partner through capacity planning, and to update the process plan by accessing the updates of manufacturing cell capacity maintained by SM from time to time, respectively. They concluded that the proposed cost function was effective for selecting optimal manufacturing partner in digital virtual manufacturing.

Zhang et al. (2003b) proposed *Concurrent Integrated Process Planning System* (CIPPS) that integrates process planning, scheduling as well as CAD. It was composed of initial planning, decision making, and detailed planning levels used in parallel for HMS environment. The authors concluded that the proposed system was capable of integrating different manufacturing functions (domains), and of responding to varying conditions of the enterprise, as well as to decrease the cost and time of processes.

Ueda et al. (2007) proposed the *Evolutionary Artificial Neural Network* (EANN) algorithm that performs the process planning and scheduling functions in parallel. EANN possessed machine learning in which each machine was capable to acquire the existing status of shop floor continuously. The machine agent made the local decision of selecting a job, and a machine to process the selected job.

Li et al. (2008a) presented a GT-based *Cooperative Process Planning and Scheduling* (CPPS) model with Pareto, Nash, and Stackelberg schemes. They solved CPPS problem by PSO, GA, and SA to minimise tardiness, makespan, manufacturing cost, and balancing machine load for a job shop. CPPS involved new order arrival and machine breakdown. They concluded that the SA algorithm performed better from a computational time viewpoint. However, the SA algorithm was dependent on parameter settings and problem size. PSO and the GA algorithms were robust to optimise the proposed problem but consumed more time to compute.

Zhanjie and Ju (2008) discussed IPPS research based on the GA and they designed effective genetic coding. AND/OR network was used to show MPP. Process routes were chosen based on the manufacturing cost, SPT dispatching rule as well as machine utilisation. Shukla et al. (2008) developed MAS-based IPPS model possessing a data-mining agent to consider the cost involved for tool-using and tool-repairing as a regularly changing value instead of a constant parameter. A component agent was developed to perform bidding for an individual feature of a job to process among all machine agents. After the assignment of features of jobs to suitable machines, the optimisation agent executed to generate an optimal schedule plan and process plan using SA and TS methods.

Zattar et al. (2008) presented MAS for IPPS based on the features negotiating protocols to reduce mean flow time and makespan in a job shop system. The proposed protocol performed real-time alteration in process plans considering alternative resources. Authors concluded that the proposed model performed better than SEA from mean flow time and makespan viewpoint. In addition, Zattar et al. (2010) reported that the reduction in makespan and flow time was due to the changes in numbers of the machine. The authors stated that consideration of processing time with setup time may lead to a poor examination of the IPPS problem.

Shrestha et al. (2008) stretched the approach presented in Sugimura et al. (2007, 2003, 2001), to compare the centralised approach and DA that modifies the process plans for IPPS in HMS. In the centralised method, the response from scheduling holons (after scheduling) was sent to the job holons and a process plan combination was generated with constraints of machining equipment. In the distributed method, the process plans were modified by job holons without considering the centralised control of scheduling holons.

Chan et al. (2006) presented an algorithm for IPPS termed as "AIS-FLC" approach that composed of *Artificial Immune System* (AIS) embedded with *Fuzzy Logic Controller* (FLC) and they considered multiple outsourced production orders with a due date and simultaneously optimised makespan performance of the system. Also, they considered an alternative sequence of operations and alternative machines for each operation in the manufacturing system of size five machines and one outsourcing station (machine). Moreover, the scheme was demonstrated as a TSP that owned the priority connections among the operations. Hybrid features of AIS-FLC, topological sort technique, and direct graph methods were used to generate a feasible sequence of operations. They stated that the proposed outsourcing approach was useful when waiting time was higher than the transportation time of jobs. Also, they claimed that the outsourcing approach decreased makespan. The extension of this approach was presented in Chan et al. (2009). They proposed *Enhanced Swift Converging Simulated Annealing* (ESCSA) algorithm for scheduling in IPPS. They compared ESCSA with SA, TS, and GA as well as TS-SA algorithms. The authors stated that ESCSA outperformed SA, TS, and GA as well as TS-SA algorithms. Also, the authors claimed that the proposed approach was a simple tool to select the outsourcing machine considering the constraints of the real shop floor.

Wang (2009) proposed a web-based sensor-driven collaborative manufacturing framework that possessed various features like real-time monitoring, adaptive process planning, dynamic scheduling, as well as remote control supported through web-based knowledge sharing. For adaptive process planning, he designed *Supervisory Planning* and *Operation Planning* layers. A client-server architecture was designed using *view-control-model* design pattern having secure session controls. The author demonstrated the feasibility of framework for a remote machining example of distributed manufacturing and he concluded that the proposed framework looked promising.

Cai et al. (2009) presented a GA-grounded cross-machine setup planning IPPS approach that possessed the feature of the flexibility of process plan modifications corresponding to the operation setups. The idea behind this approach was to use *tool accessibility analysis* for diverse machine specifications to form a machining sequence that can be assembled in a specific setup. After the setup formation, requirements from the scheduling system were assessed to generate optimal setup plans for optimisation of machine utilisation and/or makespan or cost, etc. Authors validated the proposed approach on three machines having four setups on three-axis machining configuration. They concluded that the proposed approach was suitable for machines of different designs, and they found it to be adaptive for varying requirements of setups owing to changes in machine availability. The authors claimed that the proposed approach was a strong contribution in IPPS research during *Adaptive Setup Planning* (ASP) when a process plan was combined with setups for alternative machines.

Li et al. (2010a) presented an IPPS solution for job shop environment through a mathematical model and three agents, namely, job agent, machine agent, and an optimisation agent, along with databases. They considered processing, sequential, and operational manufacturing flexibilities during process planning function, and optimised for production time. The scheduling objective was makespan. The authors concluded that the proposed approach performed rescheduling through the machine agent's negotiation with job agents and optimisation agents whenever the changes occurred at the shop floor. Lian et al. (2012) proposed population-based evolutionary algorithm, called *Imperialist Competitive Algorithm* (ICA), to optimise makespan of the IPPS problem. The solution (schedule plan and process plan) was encoded corresponding to AND/OR graph information that composed of scheduling string, operation sequence string, machines sequence string, and OR-connectors string for each job. Hence, the schedule plan and process plan for each job were generated concurrently and explored various manufacturing flexibility. The authors elaborated various steps of ICA, viz., assimilation, imperialistic competition, revolution, and elimination with an example. They conclude that ICA outperformed the SEA and GA approaches.

Lihong and Shengping (2012) proposed an improved GA and mathematical model for IPPS to optimise makespan and mean flow time performance measures, while generating process plan and schedule plan concurrently. They proposed an improved selection method to initialise the population that preserves the best process plan for each part, to assist the scheduling for achieving the possible lowest bound. They presented new genetic representation and genetic operators. The authors claimed that the proposed approach achieved good results as compared to the SEA and *modified genetic algorithm* (MGA).

Seker et al. (2013) proposed a GA and *Fuzzy Neural Network* (FNN)-based model for IPPS in parallel to handle shop floor fluctuations. They used *clustered chromosome structure* to improve the GA performance. FNN model provided a new schedule considering changes in production constraints as soon as the GA sent data. The authors concluded that FNN yielded accurate predictions during re-planning and efficient plans were generated as per the changes in production constraints. They claimed that the proposed system can be implemented without reorganising the factory setting, and process plans and schedules were generated separately in parallel.

Yu et al. (2018) proposed a DA-based *Discrete Particle Swarm Optimisation* (DPSO) algorithm for IPPS consisting of two phases, namely, (i) preplanning phase and (ii) final planning phase. In the first phase, a process network was generated for each part with consideration of static shop floor status, whereas in the second phase, the process plan and scheduling plan for each part were generated as per the current shop floor status simultaneously. The authors proposed *external archive* to avoid local convergence by keeping more than one elite solution and *mutation operator* to explore local search ability of the proposed algorithm. They verified the effectiveness and efficiency of the proposed algorithm and approach through experimentations.

2.2.4 Other Approaches

This section discusses the miscellaneous IPPS approaches that are not covered under any of the above classifications.

Liao et al. (1993) modified a CAPP system to participate in scheduling function and they named it *Computer Managed Process Planning* (CMPP). It was composed of *Process Decision Model File* (PDMF) and *Machine Tool File* (MTF). They proposed an *operation-machine indexing* (OMI) method to select the best alternative machine during process planning and to reduce the tardy jobs as well as to minimise flow time. There were two modifications, viz., (i) a conception of secondary MTF that possessed OMI data, and (ii) a software program to modify decision rules in the PDMF file. They endorse to modify the existing CAPP models for IPPS rather than establishing a fresh model from scratch.

Zijm and Kals (1995) presented an IPPS methodology for small size batch production systems to reduce lateness of jobs batch using *Critical Path Method* (CPM). It automatically chooses an operation having the largest value of lateness performance measure. They applied the CPM in *Jobplanner* software package—a *multi-resources shop floor planning and scheduling system* that was framed with *interactive scheduling mode, monitoring and control system* and *automatic scheduler* layers as well as central database interfaced with MRP system.

The authors claimed that the proposed approach reduced manufacturing lead time by 10%–30%.

Guo et al. (2009) applied PSO algorithm for IPPS to minimise makespan and job tardiness and to balance the utilisation level of machines. A re-planning method was developed to handle new production orders and machine breakdowns. The solution (sequence of operations) was coded as particles of the PSO algorithm. These particles quest for the finest solution by facilitating the optimisation capability algorithm. They stated that the performance of PSO algorithm surpassed the SA and GA from a computational efficiency viewpoint. Also, the authors concluded that PSO algorithm presented a performance improvement due to the application of crossover operator of GA.

Shao et al. (2009) proposed the combination of NLA and DA-based unique IPPS model that used simulation-based MGA to optimise makespan, as well as to stabilise the level of percentage utilisation of machines. They concluded that the presented model outperformed the traditional manufacturing system.

Sormaz et al. (2010) developed a real-time integration system using XML data representation to perform process planning and scheduling with product designing, as well as to control the functioning of FMS. They have considered two circumstances: (i) integration of CAPP and scheduling and (ii) integration of CAPP with FMS control simulation and validated the proposed model on nine selected prismatic components. They used Unigraphics© CAD software to create a features-based model and ARENA® simulation software was used to develop discrete event model of FMS. They applied rule-based method to select processes and accessibility of alternative machines corresponding to the processing time data. They stated that the presented system curtailed the progress cycle of integrating different production planning and control functions of enterprises.

Phanden et al. (2013) proposed an IPPS approach that contains features of CLA and NLA. The authors claimed that it can be readily implemented in the existing process planning and scheduling departments of any manufacturing company. It was composed of four modules, viz., PPSM, SM, *Scheduling Analysis Module* (SAM), and *Process Plan Modification Module* (PPMM). The proposed model was optimised using GA in which the job shop layout was simulated on ProModel© software to compute fitness values of solutions. They developed IPPS heuristics for makespan and mean tardiness performance measures. The authors concluded that the proposed approach was effective enough to improve system performance as compared to the hierarchical approach. Readers may refer to Chapter 2 for more details on this approach.

Zhang and Wong (2016) proposed an *Enhanced version of Ant Colony Optimisation* (E-ACO)-based IPPS approach for a job shop that considered sequence-dependent setup time and optimises for makespan fitness. The authors compared the proposed IPPS corresponding to the following scenarios: (i) considering separate setup time, (ii) including setup time into processing time, and (iii) totally overlooking setup time. They considered various setup issues such as part loading/unloading, tool changing, material transportation, and fixture preparation. The authors concluded that (i) setup time must be considered under various circumstances, (ii) separate consideration of different setup times would be a better decision, (iii) if, the setup time is less than 10% of the processing time then it could be overlooked in order to shorten run time, (iv) tool changing time can be ignored if the machines are equipped with automatic tool changing mechanism. In Zhang and Wong (2018), the authors presented E-ACO for IPPS problem evolving AND/OR network search while optimising makespan, flow time, and computational time for a job shop. They modified ACO in terms of measuring the level of convergence, introducing *node-based pheromone,* and introducing the *earliest finishing time-based* scheme in order to find heuristic desirability and formalised oriented elitist pheromone deposit scheme. The authors concluded that the proposed modifications in ACO outperformed other meta-heuristics.

Manupati et al. (2016) proposed mobile agents–based negotiation system and mathematical model for multiple objectives for IPPS in the network (integrated) manufacturing enterprise. They applied ontology through Protégé software to convert data and knowledge into an XML scheme of *Web Ontology Language* (OWL) documents. XML schema was used to transfer information through the manufacturing network for integration of model data and resources. They compared and validated the presented mobile agents–based IPPS algorithm with hybrid dynamic-DNA algorithm.

Chaudhry and Usman (2017) proposed a spreadsheet-based GA model for IPPS that can be used for general purpose to optimise all shop models and objectives without changing basic optimisation parameters setting. The authors stated that the crossover rate and a certain degree of mutation rate play an important role in the performance of GA. Also, low mutation rate was recommended for large population size and vice versa. The authors concluded that GA has the potential to make a *general-purpose real-world scheduler* using spreadsheets.

Liu et al. (2018) proposed a quantum-inspired evolutionary algorithm for IPPS in a dynamic job shop to minimise makespan. They presented a hybrid-encoding scheme that composed of three layers of numerical representation, as well as Q-bit representation based on a quantum-inspired algorithm. Also, they presented a logistic map strategy to diversify the population. They stated that the developed IPPS system was effective enough to integrate both functions easily. Bryan (2018) presented a case study on IPPS that was named *Emphasises Effective Communication through Facilitated Information and Knowledge Transfer* (EC-FIKT) for manufacturing companies of developing countries. Ba et al. (2018) proposed PSO-based IPPS model for multi-workshops and batch manufacturing environments using the equal batch splitting scheme to optimise makespan.

2.3 Analysis of Literature Review

It presents the following facts on IPPS.

- Earlier, the primary concerns of researchers were either regarding process planning or scheduling functions as these functions were considered separately in a conventional manufacturing system. In process planning function, the focus is to automate it. Scheduling takes input from the process planning function and assigns manufacturing resources to the operations defined in the process plan. Considering process planning and scheduling separately results in several problems and an unpredictable behaviour of the manufacturing system. Thus, a need is felt to integrate these two functions to cope up with the dynamic nature of the manufacturing system.

- There are three approaches to IPPS, viz., NLA, CLA, and DA. DA is the best approach for IPPS. However, DA and CLA suffer from major disadvantages such as requirements of highly capable hardware and software, dismantling and reorganising existing process planning and scheduling departments in a company, while NLA is based on static shop floor conditions.

- It has been closely observed that the MAS approach is recorded as a promising DA to develop an adaptive IPPS system. However, the agent-based system performs poorly when the levels of negotiation and the number of agents are more and vice versa.

- The search space of the IPPS problem becomes larger after merging process planning and scheduling functions. Therefore, the time taken to find the optimal solution also increases significantly. Thus, it has been observed that a hybrid approach is a viable option to find an optimal solution in a shorter time span. Consequently, a combination of meta-heuristics methods and heuristic (nature-inspired) algorithms may be helpful in solving the IPPS problem promptly.

- Most of the researchers considered a single objective to solve IPPS, such as makespan, mean flow time and tardiness, number of tardy parts, etc.

- There are several IPPS testbed problems comprising a varying number of jobs and machines that have been presented by researchers to analyse the developed approaches. Chapter 10 presents various testbed problems on IPPS.

- It has been observed that there is a need to develop an IPPS framework that can be implemented in a company with existing scheduling and planning facilities. This could be achieved by combining different integration approaches like NLA with CLA or DA in order to combine the benefits from each individual approach. This idea may help to increase the level of information exchange among the scheduling and planning departments of a company.

- Job shop manufacturing environment is an important type of shop configuration. Scheduling for a job shop is typical and NP-hard. Moreover, a close examination of the job shop scheduling problem finds quality insights in order to solve the problems confronted during the scheduling of complicated and realistic systems.

- There are two approaches to generate the schedule for the IPPS problem, viz., offline and online. Most of the researchers follow offline scheduling strategy as it plays an important role in the strategic planning of external activities like preventive maintenance and material procurement procedures. However, offline schedules may become infeasible due to some unexpected disturbances such as machine breakdown during the production process, and it may require a re-scheduling process. In online scheduling, the parts are allocated to different machines in real time by simulating the shop floor environment.

- Researchers considered the different sizes of job shop manufacturing environments in terms of the number of jobs and machines for production scheduling.

- There are various approaches for assigning due dates to the jobs such as *Processing Plus Waiting* (PPW) approach, TWK, and *Constant SlacK* (SLK) approach. However, TWK approach is most commonly used by researchers.

- There are various dispatching rules used for IPPS problem such as FCFS, SPT, LPT, SST, ECT, EDD, TWR, *Random Dispatching* (RD), *Minimum Slack Time* (MST), *Minimum Slack Time per Operation* (S/OPN), *Cost-OVER-Time* (COVERT), *Most WorK Remaining* (MWKR), and *Least WorK Remaining* (LWKR).

- Batch size in the IPPS problem could be considered in order to reduce the total completion time of a production order.

2.4 Conclusion

The sustained interest in the process planning, scheduling, and integration of these two functions has resulted in the development of many approaches as discussed in this chapter. The need for IPPS has been identified, and various fundamental methods to achieve IPPS along with their benefits and shortcomings have been discussed. Moreover, proposed architectures and methodologies, as well as results and findings of many researchers on IPPS, have been presented briefly. In the end, various facts of the developed IPPS approaches have been summarised and research directions have been suggested.

References

Aanen, E., Gaalman, G.J. and Nawijn, W.M. (1989). Planning and scheduling in an FMS. *Engineering Costs and Production Economics*, 17(1–4), 89–97.

Aldakhilallah, K.A. and Ramesh, R. (1999). Computer-Integrated Process Planning and Scheduling (CIPPS): intelligent support for product design, process planning and control. *International Journal of Production Research*, 37(3), 481–500.

Ba, L., et al. (2018). A mathematical model for multiworkshop IPPS problem in batch production. *Mathematical Problems in Engineering*, 2018, 1–16. doi:10.1155/2018/7948693

Baykasoğlu, A. and Özbakır, L. (2009). A grammatical optimization approach for integrated process planning and scheduling. *Journal of Intelligent Manufacturing*, 20(2), 211–221.

Bryan, M.E. (2018). Integration of process planning and scheduling in the manufacturing sector to enhance productivity–a case study of developing countries. Thesis of Doctor of Philosophy, University of Wolverhampton.

Cai, N., Wang, L. and Feng, H.Y. (2009). GA-based adaptive setup planning toward process planning and scheduling integration. *International Journal of Production Research*, 47(10), 2745–2766.

Chan, F.T.S., Zhang, J. and Li, P. (2001). Modelling of integrated, distributed and cooperative process planning system using an agent-based approach. *Proceedings of the Institution of Mechanical Engineers, Part B: Journal of Engineering Manufacturing*, 215, 1437–1451.

Chan, F., Kumar, V. and Tiwari, M. (2006) Optimizing the performance of an integrated process planning and scheduling problem: an AIS-FLC based approach. *Proceedings of IEEE Conference on Cybernetics and Intelligent Systems*, Bangkok, Thailand, 7–9 June, pp. 1–8.

Chan, F.T.S., Kumar, V. and Tiwari, M. (2009). The relevance of outsourcing and leagile strategies in performance optimization of an integrated process planning and scheduling model. *International Journal of Production Research*, 47(1), 119–142.

Chaudhry, I.A. and Usman, M. (2017). Integrated process planning and scheduling using genetic algorithms. *Tehnicki vjesnik—Technical Gazette*, 24(5), 1401–1409.

Cho, K.K., et al. (1998). An integrated process planning and scheduling system for block assembly in shipbuilding. *Annals of the CIRP*, 47(1), 419–422.

Choi, H. and Park, B. (2006). Integration of process planning and job shop scheduling using genetic algorithm. *Proceedings of 6th WSEAS International Conference on Simulation, Modelling and Optimization*, Lisbon, Portugal, 22–24 September, pp. 13–18.

Chryssolouris, G. and Chan, S. (1985) An integrated approach to process planning and scheduling, *Annals of the CIRP*, 34(1), 413–417.

Dong, J., Jo, H.H. and Parsaei, H.R. (1992). A feature-based dynamic process planning and scheduling. *Computer & Industrial Engineering*, 23, 141–144.

Giffler, B. and Thompson, G.L. (1960). Algorithms for solving production scheduling problems. *International Journal of Operations Research*, 8, 487–503.

Gindy, N., Saad, S. and Yue, Y. (1999). Manufacturing responsiveness through integrated process planning and scheduling. *International Journal of Production Research*, 37(11), 2399–2418.

Gu, P., Balasubramanian, S. and Norrie, D. (1997). Bidding-based process planning and scheduling in a multi-agent system. *Computer & Industrial Engineering*, 32(2), 477–496.

Guo, Y.W., et al. (2009). Optimisation of integrated process planning and scheduling using a particle swarm optimisation approach. *International Journal of Production Research*, 47(14), 3775–3796.

Haddadzade, M., Razfar, M.R. and Farahnakian, M. (2009). Integrating process planning and scheduling for prismatic parts regard to due date. *Proceedings of World Academy of Science, Engineering and Technology*, 51, 64–66.

Haddadzade, M., Razfar, M.R. and Fazel Zarandi, M.H. (2014). Integration of process planning and job shop scheduling with stochastic processing time. *The International Journal of Advanced Manufacturing Technology*, 71(1–4), 241–252.

Huang, S.S., Zhang, H.C. and Smith, M.L. (1995). A progressive approach for the integration of process planning and scheduling. *IIE Transactions*, 27, 456–464.

Jablonski, S., Reinwald, B. and Ruf, T. (1990). Integration of process planning and job shop scheduling for dynamic and adaptive manufacturing control. *Proceedings of Rensselaer's 2nd International Conference on Computer Integrated Manufacturing*, 21–23 May, pp. 444–450.

Jain, A., Jain, P. and Singh, I. (2006). An integrated scheme for process planning and scheduling in FMS. *The International Journal of Advanced Manufacturing Technology*, 30, 1111–1118.

Jain, A., et al. (2013). A review on manufacturing flexibility. *International Journal of Production Research*, 51(19), 5946–5970.

Jin, L., et al. (2016). A multi-objective memetic algorithm for integrated process planning and scheduling. *The International Journal of Advanced Manufacturing Technology*, 85(5–8), 1513–28.

Kempenaers, J., Pinte, J. and Detand, J. (1996). A collaborative process planning and scheduling system. *Advances in Engineering Software*, 25, 3–8.

Khoshnevis, B. and Chen, Q. (1991). Integration of process planning and scheduling function. *Journal of Intelligent Manufacturing*, 2(3), 165–175.

Kim, K. and Egbelu, P. (1999). Scheduling in a production environment with multiple process plans per job. *International Journal of Production Research*, 37(12), 2725–2753.

Kim, Y., Park, K. and Ko, J. (2003). A symbiotic evolutionary algorithm for the integration of process planning and job shop scheduling. *Computers & Operations Research*, 30, 1151–1171.

Kruth, J.P. and Detand, J. (1992). A CAPP system for nonlinear process plans. *CIRP Annals—Manufacturing Technology*, 41(1), 489–492.

Kumar, M. and Rajotia, S. (2002). An architecture of computer aided process-planning system integrated with scheduling using decision support system. *Proceedings of 10th International Manufacturing Conference in China (IMCC2002)*, Xiamen, China, Vol. 2, pp. 145–149.

Kumar, M. and Rajotia, S. (2003). Integration of scheduling with computer aided process planning. *Journal of Materials Processing Technology*, 138, 297–300.

Kumar, M. and Rajotia, S. (2006). Integration of process planning and scheduling in a job shop environment. *The International Journal of Advanced Manufacturing Technology*, 28(1–2), 109–116.

Lee, H. and Kim, S. (2001). Integration of process planning and scheduling using simulation based genetic algorithms. *The International Journal of Advanced Manufacturing Technology*, 18, 586–590.

Leung, C.W., et al. (2010). Integrated process planning and scheduling by an agent-based ant colony optimization. *Computer & Industrial Engineering*, 59(1), 166–180.

Li, W. and McMahon, C. (2007). A simulated annealing-based optimization approach for integrated process planning and scheduling. *International Journal of Computer Integrated Manufacturing*, 20(1), 80–95.

Li, W., et al. (2008a). Game theory-based cooperation of process planning and scheduling. *Proceedings of 12th International Conference on Computer Supported Cooperative Work in Design (CSCWD)*, Xi'an, China, pp. 841–845.

Li, X., et al. (2008b). A genetic algorithm for integration of process planning and scheduling problem. In C. Xiong, H. Liu, Y. Huang, and Y. Xiong (Eds.), *Intelligent Robotics and Applications. ICIRA 2008*. Lecture Notes in Computer Science, Vol. 5315. Berlin and Heidelberg: Springer, pp. 495–502.

Li, X., et al. (2010a). An agent-based approach for integrated process planning and scheduling. *Expert Systems with Applications*, 37, 1256–1264.

Li, X., et al. (2010b). A review on integrated process planning and scheduling. *International Journal of Manufacturing Research*, 5(2), 161–180.

Li, X., et al. (2010c). An effective hybrid algorithm for integrated process planning and scheduling. *International Journal of Production Economics*, 126(2), 289–298.

Li, X., Gao, L. and Li, W. (2012a). Application of game theory based hybrid algorithm for multi-objective integrated process planning and scheduling. *Expert Systems with Applications*, 39(1), 288–297.

Li, X., et al. (2012b). An active learning genetic algorithm for integrated process planning and scheduling. *Expert Systems with Applications*, 39(8), 6683–6691.

Lian, K., et al. (2012). Integrated process planning and scheduling using an imperialist competitive algorithm. *International Journal of Production Research*, 50(15): 4326–4343.

Liao, T.W., et al. (1993). Modification of CAPP systems for CAPP/scheduling integration. *Computer & Industrial Engineering*, 25(1–4), 203–206.

Lihong, Q. and Shengping, L. (2012). An improved genetic algorithm for integrated process planning and scheduling. *The International Journal of Advanced Manufacturing Technology*, 58(5), 727–740.

Liu, M., Yi, S. and Wen, P. (2018). Quantum-inspired hybrid algorithm for integrated process planning and scheduling. *Proceedings of the Institution of Mechanical Engineers, Part B: Journal of Engineering Manufacture*, 232(6), 1105–1122.

Mamalis, A.G., et al. (1996). On-line integration of a process planning module with production scheduling. *The International Journal of Advanced Manufacturing Technology*, 12, 330–338.

Manupati V.K., et al. (2016). Integration of process planning and scheduling using mobile-agent based approach in a networked manufacturing environment, *Computers & Industrial Engineering*, 94, 63–73.

Moon, C., et al. (2002). Integrated process planning and scheduling with minimizing total tardiness in multi-plants supply chain. *Computers & Industrial Engineering*, 43, 331–349.

Moon, C., et al. (2008). Integrated process planning and scheduling in a supply chain. *Computer & Industrial Engineering*, 54(4), 1048–1061.

Morad, N. and Zalzala, A. (1999). Genetic algorithms in integrated process planning and scheduling. *Journal of Intelligent Manufacturing*, 10, 169–179.

Naseri, M.R.A. and Afshari, A.J. (2012). A hybrid genetic algorithm for integrated process planning and scheduling problem with precedence constraints. *The International Journal of Advanced Manufacturing Technology*, 59, 273–287.

Palmer, J. (1996). A simulated annealing approach to integrated production scheduling. *Journal of Intelligent Manufacturing*, 7, 163–176.

Petrović, M., et al. (2016a). The ant lion optimization algorithm for integrated process planning and scheduling. *Applied Mechanics and Materials*, 834, 187–192.

Petrović, M., et al. (2016b). Integration of process planning and scheduling using chaotic particle swarm optimization algorithm. *Expert Systems with Applications*, 64, 569–588.

Phanden, R.K., et al. (2011). Integration of process planning and scheduling: a state-of-the-art review." *International Journal of Computer Integrated Manufacturing*, 24(6), 517–534.

Phanden, R.K., et al. (2013). An approach for integration of process planning and scheduling. *International Journal of Computer Integrated Manufacturing*, 26(4), 284–302.

Sadeh, N., et al. (1998). A blackboard architecture for integrating process planning and production scheduling. *Concurrent Engineering: Research and Applications*, 6(2), 88–100.

Saygin, C. and Kilic, S.E. (1999). Integrating flexible process plans with scheduling in flexible manufacturing systems. *The International Journal of Advanced Manufacturing Technology*, 15, 268–280.

Seker, A., Erol, S. and Botsali, R. (2013). A neuro-fuzzy model for a new hybrid integrated process planning and scheduling system. *Expert Systems with Applications*, 40(13), 5341–5351.

Shao, X., et al. (2009). Integration of process planning and scheduling: a modified genetic algorithm-based approach. *Computers and Operations Research*, 36, 2082–2096.

Shokouhi, E. (2018). Integrated multi-objective process planning and flexible job shop scheduling considering precedence constraints. *Production & Manufacturing Research*, 6(1), 61–89.

Shrestha, R., et al. (2008). A study on integration of process planning and scheduling system for holonic manufacturing with modification of process plans. *International Journal of Manufacturing Technology and Management*, 14(3–4), 359–378.

Shukla, S., Tiwari, M. and Son, Y. (2008). Bidding-based multi-agent system for integrated process planning and scheduling: a data-mining and hybrid tabu-SA algorithm-oriented approach. *International Journal of Advance Manufacturing Technology*, 38, 163–175.

Sobeyko, O. and Mönch, L. (2017). Integrated process planning and scheduling for large-scale flexible job shops using metaheuristics. *International Journal of Production Research*, 55(2), 392–409.

Sormaz, D.N., et al. (2010). Integration of product design, process planning, scheduling, and FMS control using XML data representation. *Robotics and Computer-Integrated Manufacturing*, 26(6), 583–595.

Sugimura, N., et al. (2001). Integrated process planning and scheduling in holonic manufacturing systems. *Proceedings of IEEE International Symposium on Assembly Task Planning*, Soft Research Park, Fukuoka, Japan, Vol. 4, pp. 250–254.

Sugimura, N., Shrestha, R. and Inoue, J. (2003). Integrated process planning and scheduling in holonic manufacturing systems—optimization based on shop time and machining cost. *Proceedings of 5th IEEE International Symposium on Assembly and Task Planning*, Besancon. IEEE, pp. 36–41.

Sugimura, N., et al. (2007) A study on integrated process planning and scheduling system for holonic manufacturing. In L. Wang and W. Shen (Eds.), *Process Planning and Scheduling for Distributed Manufacturing*. London: Springer, pp. 311–334.

Sundaram, R.M. and Fu, S.S. (1988). Process planning and scheduling. *Computer and Industrial Engineering*, 15, 296–307.

Tan, W. (1998) Integration of process planning and scheduling—a mathematical programming approach. PhD Dissertation, University of Southern California.

Tan, W. and Khoshnevis, B. (2000). Integration of process planning and scheduling: a review. *Journal of Intelligent Manufacturing*, 3052, 51–63.

Tonshoff, H.K., Beckendorff, U. and Andres, N. (1989). *FLEXPLAN: A Concept for Intelligent Process Planning and Scheduling*. CIRP Inter. Workshop, Germany, pp. 21–22.

Ueda, K., Fuji, N. and Inoue, R., 2007. An emergent synthesis approach to simultaneous process planning and scheduling. *Annals of CIRP*, 56(1), 463–466.

Usher, J. and Fernandes, K. (1996) Dynamic process planning – the static phase. *Journal of Materials Processing Technology*, 61, 53–58.

Wang, Y.F., et al. (2008a). An integrated approach to reactive scheduling subject to machine breakdown. *2008 IEEE International Conference on Automation and Logistics*, 1–3 September, Qingdao, China, pp. 542–547.

Wang, Y.F., Zhang, Y.F., and Fuh, J.Y.H. (2008b). A web-based integrated process planning and scheduling system. *IEEE Conference on Automation Science and Engineering*, pp. 662–667.

Wang, L. (2009). Web-based decision making for collaborative manufacturing. *International Journal of Computer Integrated Manufacturing*, 22(4), 334–344.

Wang, J., et al. (2014). A graph-based ant colony optimization approach for integrated process planning and scheduling. *Chinese Journal of Chemical Engineering*, 22(7), 748–753.

Weintraub, A., et al. (1999). Scheduling with alternatives: a link between process planning and scheduling. *IIE Transactions*, 31, 1093–1102.

Wong, T.N., et al. (2006a). An agent-based negotiation approach to integrate process planning and scheduling. *International Journal of Production Research*, 44(7), 1331–1351.

Wong, T.N., et al. (2006b). Integrated process planning and scheduling/rescheduling-an agent-based approach. *International Journal of Production Research*, 44(18–19), 3627–3655.

Wu, S., Fuh, J. and Nee, A. (2002). Concurrent process planning and scheduling in distributed virtual manufacturing. *IIE Transactions*, 34, 77–89.

Xia, H., Li, X. and Gao, L. (2016) A hybrid genetic algorithm with variable neighborhood search for dynamic integrated process planning and scheduling. *Computers & Industrial Engineering*, 102, 99–112.

Yang, Y., Parsaei, H. and Leep, H. (2001). A prototype of a feature-based multiple-alternative process planning system with scheduling verification. *Computers & Industrial Engineering*, 39, 109–124.

Yu, M.R., Yang, B. and Chen, Y. (2018). Dynamic integration of process planning and scheduling using a discrete particle swarm optimization algorithm. *Advances in Production Engineering & Management*, 13(3), 279–296.

Zattar, I., et al. (2008). Integration between process planning and scheduling using feature-based time-extended negotiation protocols in a multi agent system. *International Journal of Services Operations and Informatics*, 3(1), 71–89.

Zattar, I., et al. (2010). A multi-agent system for the integration of process planning and scheduling using operation-based time-extended negotiation protocols. *International Journal of Computer Integrated Manufacturing*, 23(5), 441–452.

Zhang, H.C. and Merchant, M.E. (1993). IPPM—A prototype to integrate process planning and job shop scheduling functions. *CIRP Annals—Manufacturing Technology*, 42(1), 513–518.

Zhang, Y., Saravanan, A. and Fuh, J. (2003a). Integration of process planning and scheduling by exploring the flexibility of process planning. *International Journal of Production Research*, 41(3), 611–628.

Zhang, J., et al. (2003b). A holonic architecture of the concurrent integrated process planning system. *Journal of Materials Processing Technology*, 139, 267–272.

Zhang, S. and Wong, T.N. (2016). Studying the impact of sequence-dependent set-up times in integrated process planning and scheduling with E-ACO heuristic. *International Journal of Production Research*, 54(16), 4815–4838.

Zhang, Z., et al. (2016). A method for minimising the energy consumption of machining system: Integration of process planning and scheduling. *Journal of Cleaner Production*, 137, 1647–1662.

Zhang, S. and Wong, T.N. (2018). Integrated process planning and scheduling: an enhanced ant colony optimization heuristic with parameter tuning. *Journal of Intelligent Manufacturing*, 29(3), 585–601.

Zhanjie, W. and Ju, T. (2008). The research about integration of process planning and production scheduling based on genetic algorithm. *International Conference on Computer Science and Software Engineering*, 1, 9–12.

Zhao, F., et al. (2004). A genetic algorithm based approach for integration of process planning and production scheduling. *Proceedings of International Conference on Intelligence Mechanical and Automotive*, China, pp. 483–488.

Zhao, F., et al. (2006). Integration of process planning and production scheduling based on a hybrid PSO and SA algorithm. *Proceeding of IEEE International Conference on Mechanical and Automotive*, Luoyang, China, 25–28 June.

Zhao, F., et al. (2010). A hybrid particle swarm optimisation algorithm and fuzzy logic for process planning and production scheduling integration in holonic manufacturing systems. *International Journal of Computer Integrated Manufacturing*, 23(1), 20–39.

Zijm, W.H.M. and Kals, H.I.J. (1995). The integration of process planning and shop floor scheduling in small batch part manufacturing. *CIRP Annals—Manufacturing Technology*, 44(1), 429–432.

3

A Case Study on Optimisation of Integrated Process Planning and Scheduling Functions Using Simulation-Based Genetic Algorithm and Heuristic for Makespan Performance Measurement

Rakesh Kumar Phanden

Amity University Uttar Pradesh

Ajai Jain

National Institute of Technology Kurukshetra

CONTENTS

3.1 Introduction 50
3.2 Problem Formulation 50
3.3 Selection of Job Shop Configuration and Production Order Details 51
3.4 Scheduling Method and Assumptions..52
3.5 IPPS Methodology 53
     3.5.1 PPSM..54
     3.5.2 SM...59
     3.5.3 SAM 59
     3.5.4 PPMM 62
3.6 SM Implementation 71
     3.6.1 Encoding Scheme...71
     3.6.2 Initialisation...73
     3.6.3 Fitness Function 73
     3.6.4 Reproduction 76
             3.6.4.1 Selection Operation...76
             3.6.4.2 Crossover...77
             3.6.4.3 Mutation 80
     3.6.5 Repairing Scheme 80
     3.6.6 Elitism..82

3.6.7   Termination Criterion 82
3.6.8   Restart Scheme 82
3.7   Case Study 84
3.8   Comparison of Formalised IPPS Methodology With
        Conventional Manufacturing Methodology 91
3.9   Epilogue 92
References 92

3.1 Introduction

The previous chapter was dedicated to a literature review on Integrated Process Planning and Scheduling (IPPS) approaches. Based on the analysis presented in Section 2.3, this chapter presents an IPPS approach that possesses the shared features of non linear approach (NLA) and closed loop approach (CLA). The following sections present the details of the problem formulation and assumptions, the objectives, the different modules of a formalised IPPS approach, a stepwise explanation of heuristic on makespan performance measure, manufacturing environment and production order details, results and comparison of IPPS with the hierarchical approach.

3.2 Problem Formulation

In this work, a job shop has been considered to process a small batch of multiple part types. A job shop is taken as a target manufacturing system as it is a suitable manufacturing strategy for present-day manufacturing environments of multi-products production orders. In this work, the IPPS problem can be expressed as: *given that there is a production order consisting of several part types, with several jobs of each part type in the given production order, and if Multiple Process Plans (MPP) for each part type are known in advance for a given job shop environment, the objective is to formalise an integration methodology by combining NLA and CLA and to assess its performance by implementing it in the job shop manufacturing system. The formalised IPPS method should be such that it can be implemented in a company with existing process planning and scheduling departments, and in which MPP of each part type remains available during job shop scheduling.*

Hence, the stated IPPS problem could be divided into the following investigation aspects:

- Formalisation of IPPS methodology for a job shop manufacturing system.
- Development of a heuristic algorithm for the formalised IPPS methodology.
- Implementation of the formalised IPPS methodology for a job shop manufacturing system.
- Assessment of the performance of the formalised IPPS methodology.
- Comparison of the formalised IPPS methodology with conventional manufacturing scenario in which both functions are carried out separately and only one process plan per part type is available to the scheduling department.

3.3 Selection of Job Shop Configuration and Production Order Details

In the present work, a job shop is taken as a target manufacturing system. It has been identified that an enormous number of companies (of small to medium sizes) are operating at job shop configuration and it is a suitable manufacturing strategy for present-day manufacturing requirements. Moreover, an investigation into the job shop scheduling environment yields important findings regarding the scheduling problems confronted during the scheduling of highly complex and real systems. The selection of production order details as well as job shop configuration goes on simultaneously.

Kim et al. (2003) presents testbed problems for IPPS. These testbed problems of varying job sizes are available at http://syslab.chonnam.ac.kr/links/data-pp&s.doc (Accessed on 7th November 2011). It presents a MPP of 18 jobs to be processed on a 15 machines job shop manufacturing environment. In the present work, MPP of jobs, as well as production order, are taken from Kim et al. (2003). The present work considers testbed problem number one of Kim (2003). The production quantity of each part type is generated randomly between 10 and 50.

The part flow in the job shop is described as follows. (i) The production order is received at the incoming location. (ii) As per the process plan, the part (say raw material) is first stacked on a machine or in the queue for its initial operation. (iii) It is processed on the machine if the machine is idle, otherwise the part enters the queue and waits for processing. (iv) As soon as the first operation is completed, it follows the next machine according to the predefined process plan for the next operation. (v) As per the process plan, the part is then moved to subsequent machines. Finally, the part (i.e., the finished part) leaves the system after finishing its machining operations.

3.4 Scheduling Method and Assumptions

There are various meta-heuristics such as tabu search, cuckoos search, variable neighbourhood search, ant colony and particle swarm optimisation, simulated annealing, memetic algorithm, genetic algorithm (GA), and many more have been utilised by researchers to optimise the job shop scheduling and IPPS problems. It has been noticed that amidst the availability of different optimisation techniques, GA is popular and famous due to its well-acknowledged optimising capability for the category of combinatorial optimisation problems. A literature review clearly reveals that GA is a promising optimisation technique to optimise the formalised IPPS problem. The present work plans to utilise simulation-based GA for scheduling. In GA, evaluation of fitness function (i.e., for makespan performance measure) is calculated through ProModel® simulation software. Simulation provides the nearby and genuine performance of a system as compared to the mathematical functions.

There are various GA operators such as selection, crossover, and mutation, that are needed for the proper functioning of GA. A literature review reveals that various researchers used different operators. The recommendations of researchers, as well as the fundamentals of GA, are kept in mind while formulating the proposed IPPS approach. Also, it has been planned to use a restart scheme to avoid the early convergence of GA (Ruiz and Maroto, 2006).

The real job shop manufacturing environment is quite complex and including all details of the system configuration into consideration, it increases the modelling complexity and needs simplifications. Thus, subsequent assumptions are considered in the present work for the purpose of simplification of modelling and analysis procedures as suggested by Kim et al. (2003) and Jain et al. (2006).

- All jobs/machines are independent of each other and available at the commencement of processing.
- MPPs of each part type are available in advance.
- Operation time of each part type is deterministic and known in advance.
- A machine can perform one operation at a time.
- Part and operation impound is forbidden.
- Infinite buffer volume is considered just ahead of a machine and the parts must arrive in the queue (buffer) before processing.
- A fixed transportation time (3 min) is considered among the machines.
- *Shortest Processing Time* (SPT) is used to select a job from the queue of a machine and if there is a tie, *First-Come-First-Serve* (FCFS) dispatching rule is applied.

3.5 IPPS Methodology

The scheme for IPPS adopted in the present work is shown in Figure 3.1. It possesses features of NLA as well as CLA. It utilises a simulation-based GA optimisation method as well as heuristics and considers MPP involved in job shop scheduling setting. In this work, the heuristic algorithm for makespan performance measure is developed for the formalised IPPS approach.

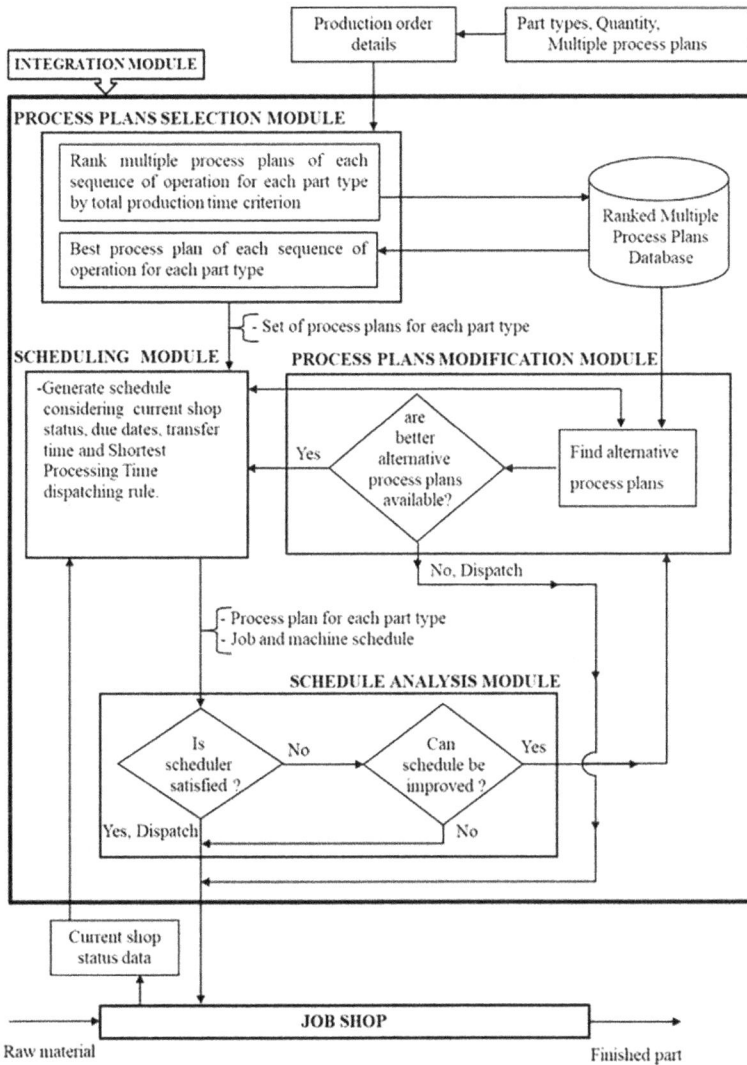

FIGURE 3.1
Proposed IPPS methodology (Phanden et al., 2013).

The formalised methodology consists of four modules, namely, *Process Plans Selection Module* (PPSM), *Scheduling Module* (SM), *Schedule Analysis Module* (SAM), and *Process Plans Modification Module* (PPMM). PPSM and SM follow NLA and SAM as well as PPMM follow the principle of CLA. ProModel® simulation software is utilised in SM. Following sections describe the details of these modules.

3.5.1 PPSM

This is the first module of the formalised IPPS. This module selects a set of best process plans for each part type of a production order. Once a production order and the MPPs of each part type of a production order are received, PPSM is invoked. It ranks the MPPs based on the minimum *Total Production Time* (TPT) criterion for each *SeQuence of operation* (SQ) for a part type and selects the best process plan corresponding to each SQ. During ranking, if a tie occurs, any one of the process plans is selected randomly. TPT is taken as a criterion because in a manufacturing system the shop time of a job includes (i) *Total Operation processing Time* (TOT), (ii) *Total Transportation Time* (TTT) and, (iii) total waiting time at queues (input and output) of the machine. Whereas, the production schedule of jobs affects the waiting time at queues of machine and the process plan of jobs affects TOT and TTT (Jain et al., 2006). Thus, for each part type, MPPs corresponding to each SQ are arranged in the decreasing order of their priority based on the minimum TPT criterion. The present work limits the maximum number of MPPs to 20 for each SQ of a given part type since a limited amount of flexibility has been suggested in the process plans to improve the system performance. Thus, for a given part type, if the number of process plans for a SQ are more than 20, then the first 20 ranked MPPs are selected. These ranked MPPs are stored in *Ranked MPP Database* and the best process plan corresponding to each SQ for a given part type is selected. The selected process plans for each part type will be ready to use during scheduling activity. Therefore, this availability of MPP safeguards a viable substitute to tackle the shop floor interruptions like machine breakdown.

It is important to mention that the number of MPPs available for each part type during scheduling depends on the number of sequences of operations for that part type of the production order. PPSM is invoked for all part types of the production order. Thus, the output of PPSM is a set of process plans for each part type of the production order.

The working of the PPSM is further explained through an example part type 2 (i.e., job number 2 from Kim et al., 2003) that has four sequences of operation as shown in Tables 3.1–3.4, respectively. The available MPPs for each operation sequence number of the example part type are shown in column (2) and TOT in column (3) of Tables 3.1–3.4, respectively. TTT (column 4) is computed for all MPPs of the example part type of production order assuming three units of transportation times between two consecutive

TABLE 3.1

Working of PPSM for an Example Part Type – First Sequence of Operation

S. No.	Multiple Process Plans for First Sequence of Operation $\{M_i\,(T_i)\}$	TOT	TTT	TPT	Ranking	Selected Process Plan $\{M_i\,(T_i)\}$
(1)	(2)	(3)	(4)	(5)	(6)	(7)
1	**5(10)-9(10)-1(33)-5(38)**	91	15	106	1	5(10)-9(10)-1(33)-5(38)
2	**5(10)-9(10)-1(76)**	96	12	108	2	
3	**14(13)-9(10)-1(33)-5(38)**	94	15	109	3	
4	**5(10)-12(13)-1(33)-5(38)**	94	15	109	4	
5	**5(10)-6(14)-1(33)-5(38)**	95	14	110	5	
6	**14(13)-9(10)-1(76)**	99	12	111	6	
7	**5(10)-12(13)-1(76)**	99	12	111	7	
8	**8(16)-9(10)-1(33)-5(38)**	97	15	112	8	
9	**5(10)-6(14)-1(76)**	100	12	112	9	
10	**14(13)-6(14)-1(33)-5(38)**	98	15	113	10	
11	**5(10)-9(10)-7(40)-5(38)**	98	15	113	11	
12	**8(16)-9(10)-1(76)**	102	12	114	12	
13	**8(16)-12(13)-1(33)-5(38)**	100	15	115	13	
14	**14(13)-6(14)-1(76)**	103	12	115	14	
15	**8(16)-6(14)-1(33)-5(38)**	101	15	116	15	
16	**14(13)-9(10)-7(40)-5(38)**	101	15	116	16	
17	**5(10)-12(13)-7(40)-5(38)**	101	15	116	17	
18	**5(10)-9(10)-11(43)-5(38)**	101	15	116	18	
19	**5(10)-7(20)-1(33)-5(38)**	101	15	116	19	
20	5(10)-6(14)-7(40)-5(38)	102	15	117	20	
21	5(10)-9(10)-7(40)-1(42)	102	15	117	21	
22	8(16)-12(13)-1(76)	105	12	117	22	
23	8(16)-6(14)-1(76)	106	12	118	23	
24	5(10)-7(20)-1(76)	106	12	118	24	
25	8(16)-9(10)-7(40)-5(38)	104	15	119	25	
26	14(13)-9(10)-11(43)-5(38)	104	15	119	26	
27	14(13)-9(10)-11(43)-5(38)	104	15	119	27	
28	14(13)-9(10)-7(40)-1(42)	105	15	120	28	
29	14(13)-6(14)-7(40)-5(38)	105	15	120	29	
30	5(10)-9(10)-11(43)-1(42)	105	15	120	30	
31	5(10)-6(14)-11(43)-5(38)	105	15	120	31	

$M_i\,(T_i)$, Machine number (Operation Processing Time); TOT, Total Operation Processing Time; TTT, Total Transportation Time; TPT, Total Production Time; Process Plans in bold are stored in *Ranked MPP Database*.

TABLE 3.2

Working of PPSM for an Example Part Type – Second Sequence of Operation

S. No.	Multiple Process Plans for Second Sequence of Operation {M_i (T_i)}	TOT	TTT	TPT	Ranking	Selected Process Plan {M_i (T_i)}
(1)	(2)	(3)	(4)	(5)	(6)	(7)
1	5(10)-8(6)-4(40)-5(38)	94	15	109	1	5(10)-8(6)-4(40)-5(38)
2	8(22)-4(40)-5(38)	100	12	112	2	
3	5(10)-8(6)-4(40)-1(42)	98	15	113	3	
4	5(10)-15(7)-4(40)-1(42)	99	15	114	4	
5	5(10)-9(8)-4(40)-1(42)	100	15	115	5	
6	14(13)-8(6)-4(40)-1(42)	101	15	116	6	
7	8(22)-4(40)-1(42)	104	12	116	7	
8	14(13)-15(7)-4(40)-1(42)	102	15	117	8	
9	8(16)-9(8)-4(40)-1(42)	106	15	121	9	

TABLE 3.3

Working of PPSM for an Example Part Type – Third Sequence of Operation

S. No	Multiple Process Plans for Third Sequence of Operation {M_i (T_i)}	TOT	TTT	TPT	Ranking	Selected Process Plan {M_i (T_i)}
(1)	(2)	(3)	(4)	(5)	(6)	(7)
1	6(25)-10(41)-14(24)-6(38)-10(37)	165	18	186	1	6(25)-10(41)-14(24)-
2	6(25)-15(44)-14(24)-6(38)-10(37)	168	18	186	2	6(38)-10(37)
3	6(25)-10(41)-14(24)-6(38)-12(40)	168	18	186	3	
4	6(25)-10(41)-14(24)-11(42)-10(37)	169	18	187	4	
5	11(33)-10(41)-15(20)-6(38)-10(37)	169	18	187	5	
6	15(30)-10(41)-14(24)-6(38)-10(37)	170	18	188	6	
7	6(25)-10(41)-15(30)-6(38)-10(37)	171	18	189	7	
8	6(25)-15(74)-6(38)-10(37)	174	15	189	8	
9	6(25)-10(41)-14(24)-11(42)-12(40)	172	18	190	9	
10	6(25)-15(44)-14(24)-11(42)-10(37)	172	18	190	10	
11	11(33)-10(41)-14(24)-6(38)-10(37)	173	18	191	11	
12	15(30)-15(44)-14(24)-6(38)-10(37)	173	18	191	12	
13	15(30)-10(41)-14(24)-6(38)-12(40)	173	18	191	13	
14	15(74)-14(24)-6(38)-12(40)	176	15	191	14	
15	15(30)-10(41)-14(24)-11(42)-10(37)	174	18	192	15	
16	6(25)-10(41)-15(30)-6(38)-12(40)	174	18	192	16	
17	15(74)-14(24)-11(42)-10(37)	177	15	192	17	
18	6(25)-15(44)-14(24)-11(42)-12(40)	175	18	193	18	
19	6(25)-10(41)-15(30)-11(42)-10(37)	175	18	193	19	

(Continued)

TABLE 3.3 (*Continued*)

Working of PPSM for an Example Part Type – Third Sequence of Operation

S. No	Multiple Process Plans for Third Sequence of Operation {M_i (T_i)}	TOT	TTT	TPT	Ranking	Selected Process Plan {M_i (T_i)}
(1)	(2)	(3)	(4)	(5)	(6)	(7)
20	**6(25)-15(74)-11(42)-10(37)**	178	15	193	20	
21	11(33)-15(44)-14(24)-6(38)-10(37)	176	18	194	21	
22	11(33)-10(41)-14(24)-6(38)-12(40)	176	18	194	22	
23	6(25)-10(41)-11(76)-10(37)	179	15	194	23	
24	15(30)-10(41)-14(24)-11(42)-12(40)	177	18	195	24	
25	15(74)-14(24)-11(42)-12(40)	180	15	195	25	
26	6(25)-15(44)-11(34)-6(38)-10(37)	178	18	196	26	
27	6(25)-10(41)-11(34)-6(38)-12(40)	178	18	196	27	
28	6(25)-10(41)-15(30)-11(42)-12(40)	178	18	196	28	
29	6(25)-15(74)-11(42)-12(40)	181	15	196	29	
30	11(33)-15(44)-14(24)-6(38)-12(40)	179	18	197	30	
31	6(25)-10(41)-11(76)-12(40)	182	15	197	31	

Note: Bold represents first 20 ranked process plans which are stored in *Ranked MPP Database*.

TABLE 3.4

Working of PPSM for an Example Part Type – Fourth Sequence of Operation

S. No	Multiple Process Plans for Fourth Sequence of Operation {M_i (T_i)}	TOT	TTT	TPT	Ranking	Selected Process Plan {M_i (T_i)}
(1)	(2)	(3)	(4)	(5)	(6)	(7)
1	**6(25)-10(41)-2(10)-4(25)-7(39)-10(37)**	177	21	198	1	6(25)-10(41)-2(10)-
2	**6(25)-10(41)-2(10)-8(26)-7(39)-10(37)**	178	21	199	2	4(25)-7(39)-10(37)
3	**6(25)-10(41)-13(12)-4(25)-7(39)-10(37)**	179	21	200	3	
4	**6(25)-15(44)-2(10)-4(25)-7(39)-10(37)**	180	21	201	4	
5	**6(25)-10(41)-13(12)-8(26)-7(39)-10(37)**	180	21	201	5	
6	**6(25)-10(41)-2(10)-4(25)-7(39)-12(40)**	180	21	201	6	
7	**6(25)-15(44)-2(10)-8(26)-7(39)-10(37)**	181	21	202	7	
8	**6(25)-15(44)-13(12)-4(25)-7(39)-10(37)**	182	21	203	8	
9	**6(25)-10(41)-2(10)-12(30)-7(39)-10(37)**	182	21	203	9	
10	**6(25)-15(44)-13(12)-8(26)-7(39)-10(37)**	183	21	204	10	
11	**6(25)-10(41)-13(12)-12(30)-7(39)-10(37)**	184	21	205	11	
12	**6(25)-15(44)-2(10)-8(26)-7(39)-12(40)**	184	21	205	12	
13	**11(33)-10(41)-2(10)-4(25)-7(39)-10(37)**	185	21	206	13	
14	**6(25)-15(44)-2(10)-12(30)-7(39)-10(37)**	185	21	206	14	
15	**6(25)-10(41)-2(10)-12(30)-7(39)-12(40)**	185	21	206	15	
16	**6(25)-15(44)-13(12)-12(30)-7(39)-10(37)**	187	21	208	16	
17	**11(33)-15(44)-2(10)-4(25)-7(39)-10(37)**	188	21	209	17	

(*Continued*)

TABLE 3.4 (*Continued*)

Working of PPSM for an Example Part Type – Fourth Sequence of Operation

S. No	Multiple Process Plans for Fourth Sequence of Operation $\{M_i (T_i)\}$	TOT	TTT	TPT	Ranking	Selected Process Plan $\{M_i (T_i)\}$
(1)	(2)	(3)	(4)	(5)	(6)	(7)
18	**11(33)-10(41)-2(10)-4(25)-7(39)-12(40)**	188	21	209	18	
19	**6(25)-15(44)-2(10)-12(30)-7(39)-12(40)**	188	21	209	19	
20	**11(33)-15(44)-13(12)-4(25)-7(39)-10(37)**	190	21	211	20	
21	11(33)-10(41)-2(10)-12(30)-7(39)-10(37)	190	21	211	21	
22	11(33)-15(44)-2(10)-4(25)-7(39)-12(40)	191	21	212	22	
23	11(33)-15(44)-13(12)-8(26)-7(39)-10(37)	191	21	212	23	
24	11(33)-15(44)-2(10)-8(26)-7(39)-12(40)	192	21	213	24	
25	11(33)-15(44)-2(10)-12(30)-7(39)-10(37)	193	21	214	25	
26	11(33)-10(41)-2(10)-12(30)-7(39)-12(40)	193	21	214	26	
27	11(33)-15(44)-13(12)-12(30)-7(39)-10(37)	195	21	216	27	
28	11(33)-15(44)-2(10)-12(30)-7(39)-12(40)	196	21	217	28	
29	15(50)-10(41)-2(10)-4(25)-7(39)-10(37)	202	21	223	29	
30	15(94)-2(10)-4(25)-7(39)-10(37)	205	18	223	30	
31	15(50)-10(41)-2(10)-8(26)-7(39)-10(37)	203	21	224	31	

Note: Bold represents first 20 ranked process plans which are stored in *Ranked MPP Database*.

TABLE 3.5

Working of PPSM – Selected Process Plans for the Example Part Type

S. No.	Operation Sequence Number	Selected Process Plans $\{M_i (T_i)\}$
(1)	(2)	(3)
1	1st	5(10)-9(10)-1(33)-5(38)
2	2nd	5(10)-8(6)-4(40)-5(38)
3	3rd	6(25)-10(41)-14(24)-6(38)-10(37)
4	4th	6(25)-10(41)-2(10)-4(25)-7(39)-10(37)

machines. TPT (column 5) of a process plan is then determined by the sum of TOT and TTT. Ranking of all the available MPPs is done based on the minimum TPT criterion (column 6). During ranking, if more than one process plan yield the same amount of TPT then any one of the process plans is selected randomly (e.g., in Table 3.3, there are three process plans having TPT of 186 units, these are ranked randomly). The first 20 ranked process plans are stored in *Ranked MPP Database* and the best process plan for a SQ is selected from these ranked MPPs (column 7). Table 3.5 shows four selected process plans for the example part type of a production order. All the selected process plans remain available during scheduling that is carried out by SM. SM is described in the following section.

3.5.2 SM

SM activates subsequent to the PPSM module. The set of selected process plans supplied by PPSM for each part type are passed to SM. SM generates an optimal schedule of parts as well as machines considering the current shop status, including precedence relationship and availability of machines using simulation-based GA. It also selects a process plan for each part type. The accessibility of updated shop floor status is the strength of SM. Moreover, in GA approach, evaluation of fitness function for makespan performance measure is carried out using simulation. Simulation outperformed the mathematical formulation since the simulation runs for a different set of situations to find the actual performance of system. For this purpose, ProModel® Simulation software is used. The output of SM is the schedule of each part as well as a machine, along with makespan of the system. It also provides the process plan followed by each part type of a production order. It is important to mention that there may be more than one schedule for a given performance measure as the present work utilises simulation-based GA approach. After schedule generation, scheduler performs its analysis in order to find the scope of further improvement through SAM.

3.5.3 SAM

The output of SM is sent to SAM. If the scheduler is satisfied with the schedule generated by SM based on makespan of system, it will be dispatched for production, otherwise SAM analyses the schedule generated. It investigates the possibility of improving the schedule depending on schedule performance measure. Thus, the output of SAM is either the accepted schedule generated by SM, or an analysis of the schedule, which provides a platform for scope of further improvement in the schedule performance. If there are more than one schedules for a given performance measure provided by SM, the scheduler will select the schedule to be analysed. The following section discusses the detailed working steps of the SAM algorithm for makespan, when the scheduler is not satisfied with the generated schedule.

Table 3.6 provides SAM algorithm for makespan (Phanden et al., 2013). Initially, the algorithm classifies the machines as utilised (if utilisation ≥50%) and under-utilised machines (if utilisation <50%). The setting of required level of machine utilisation is formulated as an adaptive parameter. This heuristic is adaptive in nature to set the desired level of machine utilisation. The scheduler can categorise the under-utilised and utilised resources as per the shop structures. Though, in the present work, if the utilisation is equal to or more than 50%, then the machine is considered as utilised. These utilised machines are organised in the decreasing order of their percentage utilisation and stored in a utilised machines (UTIMAC) list. Similarly, the under-utilised machines are arranged in the decreasing order of their percentage utilisation and stored in a under utilised machines (UNUTIMAC) list. Now, the algorithm checks

TABLE 3.6

SAM Algorithm for Makespan

BEGIN	
Step-1	Find the percentage of machine utilisation for every machine and classify them as utilised (if utilisation $\geq 50\%$) and under-utilised machines (if utilisation $< 50\%$). (Scheduler can set the required level of machine utilisation to classify as utilised and un-utilised machines as it is generally different for different shop configurations. However, present work assumes that if utilisation is equal to or more than 50%, then the machine is utilised.)
Step-2	Prepare UTIMAC list: arrange utilised machines in decreasing order of percentage utilisation, i.e., $[M_1, M_2, M_3,..., M_m]$.
Step-3	Prepare UNUTIMAC list: arrange under-utilised machines in decreasing order of percentage utilisation of machines, i.e., $[M_a, M_b, M_c,..., M_n]$.
Step-4	If UNUTIMAC list is empty, i.e., there is no under-utilised machines, print *schedule improvement is not possible* and go to step 9.
Step-5	Prepare already checked part list (ALCPL): prepare a list of already checked parts, i.e., a list of parts which have been checked for schedule improvement. If SAM is activated the first time, then ALCPL is empty.
Step-6	Prepare OPERATION list: find average operation waiting time of each operation for each part type $[^1O_1, ^1O_2, ^1O_3..., ^1O_n, ^2O_1, ^2O_2, ^2O_3,..., ^2O_n, ^3O_1, ^3O_2, ^3O_3,..., ^3O_n]$, by arranging them in decreasing order of average operation waiting time for each part type. Remove operations of those part types from OPERATION list, which are present in ALCPL.
Step-7	If OPERATION list is empty, print *schedule improvement is not possible* and go to step 9 else go to next step.
Step-8	If OPERATION list is not empty, print *schedule improvement is possible* and go to next step.
Step-9	Stop.
END	

this list for available machines. If the UNUTIMAC list is empty, i.e., there are no under-utilised machines, the schedule cannot improve over its current status and the algorithm concludes that *schedule improvement is not possible* and stops. At this stage, the schedule will be dispatched for production.

If the UNUTIMAC list is not empty and contains any under-utilised machines, the SAM algorithm proceeds further and prepares a OPERATION list. The preparation of a OPERATION list requires average of operation waiting times of each operation for each part type and it is supplied by SM. Now, the part types are arranged in the decreasing order of their average operation waiting times. At this stage, the part types that are present in already checked part list (ALCPL) are removed from the list. Here, the algorithm checks the availability of a part type in the OPERATION list. If the OPERATION list is empty, the schedule cannot be improved over its current status and the algorithm indicates that *schedule improvement is not possible* and stops. If the OPERATION list contains any part type, it indicates that there is a scope for further improvement over the current schedule performance.

Table 3.7 shows machines numbered 1–15 (column 2) listed with their corresponding percentage utilisation (column 3). All machines are arranged in the decreasing order of percentage machine utilisation (column 4). As stated earlier, the present work assumes that if the utilisation is equal to or more than 50%, then the machine is considered as utilised. Therefore, the arranged list of machines with decreasing order of percentage of utilisation from S. No. 1 to 9 falls under the category of utilised machines, and the list from S. No. 10 to 15 under the category of under-utilised machines. Tables 3.8 and 3.9 show the UTIMAC and UNUTIMAC lists for the example schedule, respectively.

The preparation of the OPERATION list is explained with an example schedule of nine-part types in a production order that are to be processed on

TABLE 3.7

Preparation of UTIMAC and UNUTIMAC Lists for an Example Schedule

S. No.	Machine	Percentage Utilisation	Machine Order in Decreasing Percentage Utilisation	
(1)	(2)	(3)	(4)	
1	M_1	52.37	M_7	Utilised machines
2	M_2	60.36	M_3	(Utilisation $\geq$ 50%)
3	M_3	91.23	M_5	
4	M_4	38.82	M_9	
5	M_5	74.16	M_2	
6	M_6	57.71	M_6	
7	M_7	91.40	M_{10}	
8	M_8	49.41	M_{11}	
9	M_9	60.44	M_1	
10	M_{10}	54.62	M_8	Under-utilised machines
11	M_{11}	52.45	M_{12}	(Utilisation < 50%)
12	M_{12}	47.76	M_{14}	
13	M_{13}	47.33	M_{13}	
14	M_{14}	47.76	M_4	
15	M_{15}	15.63	M_{15}	

TABLE 3.8

UTIMAC List for an Example Schedule

S. No.	Machine	Percentage Utilisation
1	M_7	91.40
2	M_3	91.23
3	M_5	74.16
4	M_9	60.44
5	M_2	60.36
6	M_6	57.71
7	M_{10}	54.62
8	M_{11}	52.45
9	M_1	52.37

TABLE 3.9

UNUTIMAC List for an Example Schedule

S. No.	Machine	Percentage Utilisation
1	M_8	49.41
2	M_{12}	47.76
3	M_{14}	47.76
4	M_{13}	47.33
5	M_4	38.82
6	M_{15}	15.63

a job shop consisting of 15 machines. Table 3.10 presents the preparation of the OPERATION list. Columns (2) and (3) of Table 3.10 indicate the nine-part types and their production quantity of the production order, respectively. Column (4) indicates the number of operations in each part type and they are obtained from the associated process plan that is supplied by SM. Column (5) indicates the average waiting time for each operation of every part type. It is computed by adding the waiting times of operations of all parts of each part type and then dividing the sum by the production quantity of that part type. Operation waiting time for every part of each part type of a production order is supplied by SM using ProModel® simulation software. Column (6) indicates the operations that are arranged in the decreasing order of their average operation time. ALCPL list is indicated in column (7). Availability of part types in ALCPL list indicates that these part types have already got a chance earlier in order to improve the performance of schedule. Therefore, these part types (part types 7 and 17) should not be included in the preparation of the OPERATION list in order to improve the schedule from its current status. Column (8) indicates the OPERATION list and operation number and it is found by removing the operations of those part types that are present in ALCPL list (column 7) from the operation number indicated in column (6).

3.5.4 PPMM

The output of SAM is sent to PPMM when the scheduler is not satisfied with the generated optimal schedule and there is a scope of further improvement in the schedule performance as assessed by SAM. The PPMM assists in selecting alternative process plans of part type using operations flexibility. It selects a part type which becomes a candidate for modification of MPP. Depending upon the target (selected) part type, schedule analysis, and performance of the schedule, PPMM identifies alternative process plans of the target part type. As a side note, this stage of the formalised method integrates the process planning and scheduling functions. To create a new schedule, the identified alternative process plans are sent to SM once more. Though the MPPs of the other part types remain unaffected while generating a fresh schedule.

TABLE 3.10

Preparation of OPERATION List for an Example Schedule

(1) S. No.	(2) Part Types in an Example Production Order	(3) Production quantity	(4) Number of Operations in each Part Type	(5) Average Operation Waiting Time	(6) Operation's Number in Decreasing Order for Average Operation Waiting Time	(7) ALCPL List	(8) OPERATION List
1	1	20	6	$1O_1$	98	$2O_2$	7	$2O_2$
2				$1O_2$	76	$7O_3$	17	$4O_2$
3				$1O_3$	114	$4O_2$	—	$18O_3$
4				$1O_4$	80	$18O_3$	—	$8O_1$
5				$1O_5$	164	$17O_1$	—	$1O_6$
6				$1O_6$	569	$8O_1$	—	$2O_4$
7	2	15	4	$2O_1$	73	$1O_6$	—	$8O_3$
8				$2O_2$	1105	$2O_4$	—	$12O_2$
9				$2O_3$	161	$15O_1$	—	$15O_1$
10				$2O_4$	485	$8O_3$	—	$15O_2$
11	4	25	3	$4O_1$	255	$12O_2$	—	$4O_1$
12				$4O_2$	907	$15O_1$	—	$18O_4$
13				$4O_3$	53	$15O_2$	—	$12O_1$
14	7	34	4	$7O_1$	201	$7O_2$	—	$1O_5$
15				$7O_2$	314	$4O_1$	—	$2O_3$
16				$7O_3$	920	$18O_4$	—	$12O_4$
17				$7O_4$	219	$7O_4$	—	$1O_3$
18	8	37	3	$8O_1$	615	$7O_1$	—	$8O_2$
19				$8O_2$	114	$12O_1$	—	$1O_1$
20				$8O_3$	458	$1O_5$	—	$1O_4$

(Continued)

TABLE 3.10 (*Continued*)

Preparation of OPERATION List for an Example Schedule

(1) S. No.	(2) Part Types in an Example Production Order	(3) Production quantity	(4) Number of Operations in each Part Type		(5) Average Operation Waiting Time	(6) Operation's Number in Decreasing Order for Average Operation Waiting Time	(7) ALCPL List	(8) OPERATION List
21	12	14	5	$^{12}O_1$	192	$^{2}O_3$	—	$^{1}O_2$
22				$^{12}O_2$	439	$^{12}O_4$	—	$^{2}O_2$
23				$^{12}O_3$	36	$^{1}O_3$	—	$^{18}O_1$
24				$^{12}O_4$	126	$^{8}O_2$	—	$^{4}O_3$
25				$^{12}O_5$	53	$^{1}O_1$	—	$^{12}O_5$
26	15	18	2	$^{15}O_1$	403	$^{1}O_4$	—	$^{18}O_2$
27				$^{15}O_2$	351	$^{1}O_2$	—	$^{12}O_3$
28	17	14	2	$^{17}O_1$	834	$^{2}O_2$	—	—
29				$^{17}O_2$	462	$^{18}O_1$	—	—
30	18	16	4	$^{18}O_1$	63	$^{4}O_3$	—	—
31				$^{18}O_2$	38	$^{12}O_5$	—	—
32				$^{18}O_3$	849	$^{18}O_2$	—	—
33				$^{18}O_4$	228	$^{12}O_3$	—	—

$^{i}O_p$, jth operation number of ith part type.

Table 3.11 shows details of PPMM algorithm for makespan (Phanden et al., 2013). It supports the reduction in makespan of production order by controlling the average operation waiting time. Average operation waiting time is reduced by shifting the operations from over-utilised machines to under-utilised machines. Initially, the algorithm identifies an alternative process plan from *Ranked MPP Database* for a selected part type by comparing machines of every operation of each ranked MPP with under-utilised machines in which two or more operations can be performed on the under-utilised machine. If no such process plan is available, then the algorithm

TABLE 3.11

PPMM Algorithm for Makespan

BEGIN	
Step-1	Select a part type: select first part type to be processed from OPERATION list and assign it to *SETPART*, i.e., *SETPART-i*, corresponding to ${}^iO_j$, where ${}^iO_j$ represents *j*th operation number of *i*th part type.
Step-2	Retrieve MPP of SETPART: identify process plan ($PP_{SETPART}$) and corresponding operation sequence number ($SQ_{SETPART}$) followed by *SETPART* from SM and retrieve associated MPP of *SETPART* from *Ranked MPP Database*.
Step-3.1	Find alternative process plans of *SETPART*: compare corresponding machines of each operation of each ranked MPP with UNUTIMAC list. If no process plans with more than one operation stages that can be processed by under-utilised machine are available, then go to step 4.1 else go to next step.
Step-3.2	Find fitness value with alternative process plans: find makespan for alternative process plans of *SETPART*, keeping process plan of other part types the same as obtained in the best fitness value of earlier schedule. If any alternative process plan yields fitness value greater than best fitness value of earlier schedule than go to step 6, else go to next step.
Step-4.1	Find machine number of *SETPART*: find operation number ($OP_{SETPART-j}$) having highest average operation waiting time (from *SETPART-i*) and corresponding machine number ($M_{SETPART} = M_{(1-15)}$) of *SETPART*. Find whether $M_{SETPART}$ belongs to UNUTIMAC list or not. If it belongs to UNUTIMAC list, delete *SETPART* from OPERATION list and go to step 1 (set $i = i + 1$) else go to step 4.2.
Step-4.2	Find alternative process plans of *SETPART*: from *Ranked MPP Database*, corresponding to $SQ_{SETPART}$ and $OP_{SETPART}$ find other process plans that contain under-utilised machines keeping other operations on the same machines as contained $PP_{SETPART}$. If no process plan is there, delete *SETPART* from OPERATION list and go to step 1 else go to step 5.
Step-5	Find fitness value with alternative process plans: find makespan with alternative process plan of *SETPART*, keeping process plans of other part types the same as obtained in the best fitness value of earlier schedule. If any alternative process plan yields fitness value greater than the best fitness value of earlier schedule than go to next step, else delete *SETPART* from OPERATION list created in step 1 and go to step 1 (set $i = i + 1$).
Step-6	If better alternative process plans of *SETPART* are available then pass to SM, otherwise dispatch the schedule to the shop floor.
Step-7	Stop.
END	

searches for alternative process plans in decreasing order of priority of the selected part type so that any one of its operations can be processed by an under-utilised machine.

Further, the algorithm carries out an evaluation with alternative process plans of the selected part type by keeping the process plan of another part type unchanged using simulation-based GA approach to find the fitness value. If any of the alternative process plans of the selected part types improves the best fitness value obtained in an earlier schedule, only then it is passed to SM. The algorithm is certain that an alternative process plan of the selected part type does not reduce the best fitness value obtained in an earlier schedule. The UTIMAC, UNUTIMAC, OPERATION lists and process plans of production order, as well as the best fitness value of makespan obtained in an earlier schedule, are inputs for the PPMM algorithm for makespan measurement. PPMM selects a part type, identifies the corresponding process plans, operation sequence number and passes alternative process plans of the selected part type to SM. *Ranked MPP Database* provides MPP of the selected part type.

The working of the PPMM algorithm for makespan measure is explained by considering a production order of nine-part types and fifteen machines. Assuming the UTIMAC, UNUTIMAC and OPERATION lists provided by SAM to PPMM for the example production order are as shown in Tables 3.8–3.10, respectively, and the best fitness value obtained in an earlier schedule is 0.0795, the stepwise explanations of the PPMM algorithm is given below.

Step-1: Select a Part Type
The algorithm is initialised by selecting a part type. Initially, the first part type having the highest average operation waiting time is selected from OPERATION list. The selected part type is called *SETPART*. The OPERATION list contains a part type as well as the associated operation number to be processed and represented by "i" corresponding to ${}^iO_j$ (where ${}^iO_j$ represents jth operation number of ith part type). For example, from the OPERATION list presented in Table 3.10, at first position ${}^2O_2$ is available. It means that the second operation number of part type "2" possesses the highest average operation waiting time. Therefore, part type 2 is selected and it is assigned to variable *SETPART*, i.e., *SETPART*-2, corresponding to ${}^2O_2$.

Step-2: Retrieve MPP of Selected Part Type
After identification of a part type, the algorithm retrieves the process plan ($PP_{SETPART}$) and operation sequence number ($SQ_{SETPART}$) of *SETPART* from SAM. Based on the $PP_{SETPART}$ and $SQ_{SETPART}$, MPP of *SETPART* are retrieved from *Ranked MPP Database*.

Table 3.12 shows selected process plans of each part type of the example production order that is supplied by SAM and process plan of the selected part type 2 are shown at S. No. 2. Table 3.2 shows MPP of *SETPART*-2 for $SQ_{SETPART} - 2$ that are retrieved from *Ranked MPP Database*.

TABLE 3.12

Process Plans Followed by Each Part Type for an Example Schedule

S. No.	Part Type	Selected Process Plan {M_i (T_i)}	SQ No.	Fitness Value of Earlier Schedule (Makespan)
(1)	(2)	(3)	(4)	(6)
1	1	14(10)-15(18)-13(30)-12(25)-1(40)-8(47)	2	0.0795
2	2	5(10)-8(6)-4(40)-5(38)	2	
3	4	6(21)-12(24)-7(23)	4	
4	7	7(12)-2(31)-6(18)-7(33)	9	
5	8	10(34)-11(16)-3(45)	8	
6	12	3(22)-1(5)-9(39)-13(19)	3	
7	15	9(47)-14(50)-8(6)	4	
8	17	11(44)-5(18)-13(16)-1(24)-4(21)	8	
9	**18**	**3(8)-6(12)-2(21)-5(46)**[a]	**1**	

[a] Bold indicates the target part type, its process plan and operation sequence number in column (2), (3) and (4), respectively, for PPMM algorithm for makespan.

Step-3.1: Find Alternative Process Plans of Selected Part Type

This step identifies alternative process plans of the selected part type in which more than one operation can be performed on the under-utilised machine(s) from *Ranked MPP Database*. This stage utilises the UNUTIMAC list and ranked MPP of the selected part type.

For selected part type 2 of the example production order, column (2) of Table 3.13 shows alternative process plans considering under-utilised

TABLE 3.13

Alternative Process Plans and Associated Fitness Values of Selected Part Type-2 for an Example Schedule

S. No.	Process Plans {M_i (T_i)}	TOT	TTT	TPT	Fitness value Scenario 1	Scenario 2
(1)	(2)	(3)	(4)	(5)	(6)	(7)
1	5(10)-8(6)-4(40)-5(38)	94	15	109	*0.0812*	—
2	8(22)-4(40)-5(38)	100	12	112	*0.0833*	—
3	5(10)-8(6)-4(40)-1(42)	98	15	113	0.0730	—
4	5(10)-**15(7)**-4(40)-1(42)	99	15	114	0.0601	—
5	**14(13)**-8(6)-4(40)-1(42)	101	15	116	*0.0860*	—
6	8(22)-4(40)-1(42)	104	12	116	0.0659	—
7	**14(13)**-**15(7)**-4(40)-1(42)	102	15	117	0.0791	—
8	8(16)-9(8)-4(40)-1(42)	106	15	121	*0.0801*	—

Note: Boldface indicates the under-utilised machines; Bold and italic face indicate that it is greater than the value obtained in an earlier schedule.

machines and ranked MPP is shown in Table 3.2. If there is no such process plan available, then the algorithm shifts to step 4.1, else it moves to the following step to find the fitness value.

Step-3.2: Find Fitness Value With Alternative Process Plans
This step identifies the fitness value of the scheduling objective function (makespan) for the production order. During fitness value evaluation, alternative process plans of the selected part type, as identified in step 3.1, are taken into consideration. However, process plans of other part types of production order remain unchanged. This step ensures that the alternative process plans of the selected part type are contributing in reducing makespan of the production order. In this step, PPMM interacts with SM in order to obtain the fitness value. The fitness value with alternative process plans of the selected part type is compared with the best fitness value obtained in an earlier schedule. If any of the alternative process plans yields a fitness value greater than that obtained in the earlier schedule, the algorithm switches over to step 6. Table 3.13 shows the alternative process plans of $SETPART-2$ and the corresponding fitness values (column 6 and scenario 1). *Scenario 1* indicates that one or more alternative process plans contain more than one under-utilised machine that yield a fitness value greater than the previous schedule; *Scenario 2* indicates that one or more alternative process plans contain one under-utilised machine that yield a fitness value greater than previous schedule plans are available and corresponding fitness values are calculated keeping the process plans of the other part types the same as obtained in an earlier schedule.

Step-4.1: Find Machine Number of Selected Part Type
This step is executed on failure of either step 3.1 or 3.2, i.e., there is neither a process plan having more than one processing stage that can be processed by an under-utilised machine, nor a process plan that yields a fitness value greater than the best fitness value obtained in an earlier schedule.

Here, in order to find a machine number of selected part type ($M_{SETPART}$), the process plan of selected part type ($PP_{SETPART}$) and operation number ($OP_{SETPART}$) of the selected part type are utilised from $SETPART-i$ corresponding to ^iO_j as obtained in step 1. The algorithm ensures that $M_{SETPART}$ is present in the UTIMAC list. If it belongs to the UNUTIMAC list, the $SETPART$ is deleted from the OPERATION list and the algorithm goes back to step 1 in order to select the next part type from the OPERATION list. This process is repeated until $M_{SETPART}$ does not belong to the UNUTIMAC list. If $M_{SETPART}$ belongs to the UTIMAC list at any stage, the algorithm switches over to the next step 4.2.

In order to explain the working of steps 4.1–5, it is assumed that the selected part type discussed in step 1–3.2 does not yield a fitness value greater than that obtained in the earlier schedule (column 7 and scenario 2 of Table 3.13). Thus, the machine number is selected from $SETPART-2$ corresponding to $^2O_2$ and $PP_{SETPART} - \{5(10)\text{-}8(6)\text{-}4(40)\text{-}5(38)\}$ and assign it to $M_{SETPART}\text{-}M_{(1-15)} - M_8$.

Table 3.9 reveals that M_8 is present in the UNUTIMAC list. Therefore, the present *SETPART-2* is deleted from the OPERATION list in Table 3.10 (column 8). Now the algorithm switches back to step 1 and selects the next part type from the OPERATION list as *SETPART-4* corresponding to $^4O_2$. At the present stage, it contains $M_{SETPART}$-M_{12} corresponding to $PP_{SETPART}$ − {6(21)-12(24)-7(23)} (column 3 of Table 3.12) and $OP_{SETPART}$-2 of *SETPART*. However, machine number 12 is also present in the UNUTIMAC list (Table 3.9) and the *SETPART-4* is deleted again from the OPERATION list and the algorithm goes back to step 1 again. Now, next *SETPART-18*, corresponding to $^{18}O_3$ is selected from the OPERATION list in Table 3.10. The present *SETPART* contains $M_{SETPART}$-M_2, corresponding to $PP_{SETPART}$ − {3(8)-6(12)-2(21)-5(46)} (column 3 and Table 3.12) and $OP_{SETPART}$-3. The present $M_{SETPART}$ does not belong to the UNUTIMAC list. Thus, $OP_{SETPART}$-3 is passed to next step 4.2 for this example.

Step-4.2: Find Alternative Process Plans of Selected Part Type
This step identifies alternative process plans of the selected part type corresponding to $SQ_{SETPART}$, $OP_{SETPART}$, and $M_{SETPART}$ as identified in step 4.1. The algorithm searches *Ranked MPP Database* corresponding to the operation sequence number of the process plans of selected part type and identifies alternative process plans which contain under-utilised machine at $OP_{SETPART}$ keeping the other operations on the same machines as specified in $PP_{SETPART}$. Here, the algorithm utilises operations flexibility and replaces the followed process plan of the selected part type with alternative process plans of the selected part type.

For the example schedule, corresponding to *SETPART-18*, $SQ_{SETPART}$-1, $OP_{SETPART}$-3, $M_{SETPART}$ − M_2 and $PP_{SETPART}$-{3(8)-6(12)-2(21)-5(46)}, Table 3.14 shows MPPs that are retrieved from *Ranked MPP Database*. Table 3.8 presents

TABLE 3.14

MPP of Target Part Type-18 (First Sequence of Operation) for an Example Schedule

S. No.	Multiple Process Plans {M_i (T_i)}	TOT	TTT	TPT
1	3(8)-6(12)-2(21)-5(46)	87	15	102
2	3(8)-8(13)-2(21)-5(46)	88	15	103
3	3(8)-6(12)-2(21)-8(47)	88	15	103
4	3(8)-8(13)-2(21)-8(47)	89	15	104
5	8(26)-2(21)-5(46)	93	12	105
6	3(8)-5(16)-2(21)-5(46)	91	15	106
7	**3(8)-6(12)-8(25)-5(46)**[a]	94	15	106
8	3(8)-5(16)-2(21)-8(47)	92	15	107
9	8(13)-6(12)-2(21)-5(46)	92	15	107
10	8(13)-6(12)-2(21)-8(47)	93	15	108
11	8(13)-5(16)-2(21)-5(46)	96	15	111
12	8(13)-5(16)-2(21)-8(47)	97	15	112

[a] Bold indicates process plan selected in step 4.2 of PPMM algorithm for makespan.

the UNUTIMAC list that contains M_8, M_{12}, M_{14}, M_{13}, M_4 and M_{15}. Table 3.14 indicates that there is one process plan available as shown in S. No. 7, i.e., {3(8)-6(12)-8(25)-5(46)}, where $M_{SETPART} - M_2$ can be replaced by an under-utilised machine (M_8) at $OP_{SETPART}$-3. Hence, this alternative process plan is passed to step 5.

If there is no alternative process plan available, the algorithm goes back to step 1.

Step-5: Find Fitness Values With Alternative Process Plans
This step is similar to step 3.2 and finds the fitness value of the scheduling objective (makespan) of a production order. During fitness value evaluation, alternative process plans of the selected part type as identified in step 4.2 are taken into consideration. However, process plans of other part types of a pro-duction order remain unchanged. As described earlier in step 3.2, this step ensures that alternative process plans of the selected part type are assisting in reducing makespan of a production order. In this step, PPMM interacts with SM in order to obtain fitness value. The fitness value with alternative process plans of the selected part type is compared with the best fitness value obtained in the earlier schedule. If any of the alternative processes plans yields fitness value greater than that obtained in an earlier schedule, then the algorithm switches to step 6, else it goes back to step 1.

For example, corresponding to alternative process plan, i.e., {3(8)-6(12)-8(25)-5(46)} of *SETPART-18*, fitness value of makespan is 0.0894, keeping pro-cess plan of other part types the same as obtained in an earlier schedule. Thus, the alternative process plan yields fitness value greater than the best fitness value (0.0795) of earlier schedule and the algorithm switches over to step 6.

If none of the alternative process plans yields a fitness value greater than that obtained in an earlier schedule, the selected part type is deleted from the OPERATION list and the algorithm goes back to step 1.

Step-6: Pass Alternative Process Plans to SM
This step sends alternative process plans of the selected part type to SM. The alternative process plans of selected part type are obtained either in step 3.2 or step 5 of the algorithm. If alternative process plans are not available of the target part type, then this step dispatches the current schedule to the shop floor.

For the example production order, if the current step is reached either through step 3.2 or step 5, then all the alternative process plans of selected part type are passed to SM for further evaluation. Moreover, in step 5, the alternative process plan of *SETPART-18* is {3(8)-6(12)-8(25)-5(46)} at S. No. 7 column (2) of Table 3.14. Hence, all MPPs of the selected part type 18 as shown in Table 3.14 are passed to SM.

Step 3.2 reveals that *SETPART-2* contains eight alternative process plans mentioned in S. No. 1–8 of Table 3.13. However, all nine MPPs of the selected part type 2 as shown in column (2) of Table 3.2 are passed to SM for further evaluation.

Step-7: Stop
At this step, the algorithm stops and ends.

The revised optimal schedule generated by SM is again sent to SAM, which performs an analysis of the revised schedule. This process continues until the scheduler is satisfied with the schedule, or if the schedule cannot be improved from its current status furthermore. In this iterative manner, integration between process planning and scheduling is achieved. The outcome of the above approach will be an optimal schedule of the job shop along with the process plan followed by each part type.

The above integration methodology possesses features of NLA as well as CLA. PPSM works on the principle of NLA, whereas SAM and PPMM work on the principle of CLA. PPSM restricts the number of process plans for each part type. It utilises the results available from the literature that a limited amount of flexibility is advantageous and assists the scheduler in improving system performance (Chryssolouris, 2005). SAM analyses the generated schedule so that PPMM can identify alternative process plans for the target (selected) part type. Thus, SAM and PPMM assist in increasing the information exchange between PPSM and SM. The decision maker for the schedule is generally a human scheduler. The details of SM are discussed in the following section.

3.6 SM Implementation

This section presents details of SM implementation. It utilises a simulation-based GA approach for the generation of an optimal schedule. Figure 3.2 shows the flowchart of SM implementation methodology. As the formalised methodology utilises GA and simulation, the various factors involved in implementation are discussed below.

3.6.1 Encoding Scheme

It is the founding step of GA in which a chromosome is designed corresponding to the viable solution of the problem, and optimisation parameters are placed inside a chromosome structure. The encoded solution is called a genotype and the chromosomes are named as a string or an individual. Different structures have been presented in literature in order to encode job shop scheduling problem (Chakraborty, 2010). The present work utilises an indirect approach for solution encoding, i.e., the solution is not encoded in the form of a schedule. Here, a gene in a chromosome is divided into two parts. The first part of a gene represents a process plan number (an integer value) and second part represents operation sequence number (another integer value) of a part type. Each gene position in the chromosome (i.e., locus)

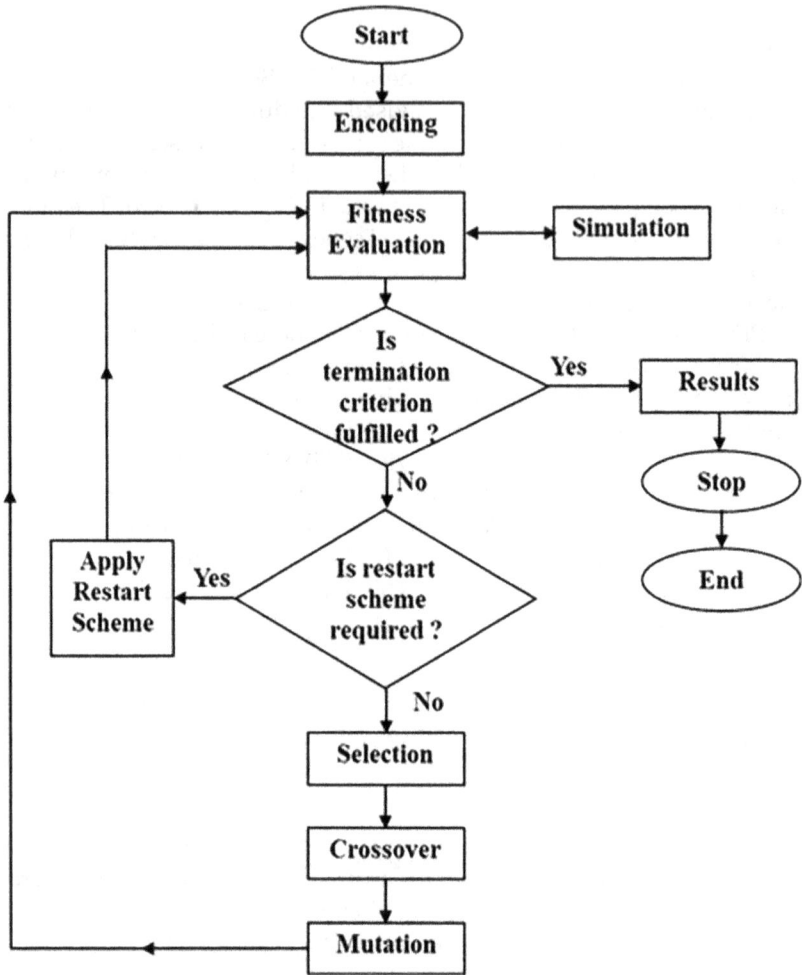

FIGURE 3.2
Scheduling methodology based on GA and simulation (Phanden et al., 2013).

is in a fixed sequence to represent associate process plan for a specific part type of production order. Each part type can be processed by any of its given MPPs from any of the associated operation sequence number. The length of the chromosome accommodates a specified number of bits (genes) and it is dependent on the number of part types in a production order. Figure 3.3 presents the chromosome structure for an example production order consisting of six-part types. In order to construct a chromosome, each gene of the chromosomes is initialised randomly from available MPP and associated SQ.

The encoding scheme adopted in the present work is explained with example part type 2 and its MPP for a first, second, third, and fourth SQs are shown in Tables 3.1–3.4, respectively. Tables 3.1–3.4 indicate the total number

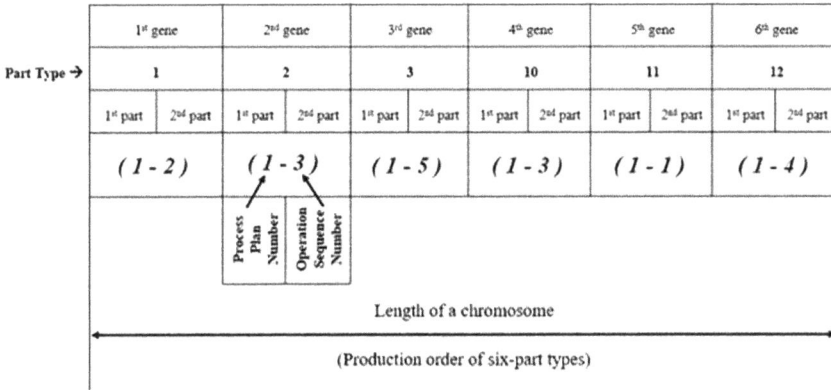

FIGURE 3.3
Chromosome structure for an example production order.

of process plans as 20, 9, 20, and 20 for a first, second, third, and fourth SQ, respectively. This information can be encoded in a gene at the second position of chromosome as shown in Figure 3.3. Here, one represent processing of part type 2 by following its 1st process plan and three represents that process plan belongs to 3rd SQ. Similarly, the information of other part types 1, 3, 10, 11, and 12 is encoded in gene number 1, 3, 4, 5, and 6, respectively. Table 3.15 presents a production order comprising six-part types 1, 2, 3, 10, 11, and 12 having two, four, five, three, three, and five sequences of operation, respectively. Column (4) presents a representation of the process plans of each part type of a production order according to the gene structure. Column (5) presents a randomly selected gene in order to create the initial population of chromosomes for the first iteration of integration methodology. During the first iteration, after execution of PPMM, MPP of the target (selected) part type (say part type 3) are modified and column (6) represents the genes for the second iteration. It shows that for the second iteration of integration methodology, the representation of MPP for 2nd SQ of part type 3 is modified.

3.6.2 Initialisation

Here, a population of specific size (*pop_size*) is evolved and termed as initial population of GA. This population possesses a desirable quantity of solutions. In general, the initial population emerges randomly (Deb, 2006). In this work, population has been generated randomly with ten number of solutions.

3.6.3 Fitness Function

A chromosome possesses a fitness value that can be called its evaluation or fitness function value. In fact, it is the objective function of GA to evaluate the performance of the algorithm. In general, the GA is most fit to solve the

TABLE 3.15

Representation of MPP for an Example Production Order of Six Part Types

Part Types	SQ No.	No. of MPP	Representation of Process Plans According to Gene Structure	Genes in 1st Iteration	Genes in 2nd Iteration
(1)	(2)	(3)	(4)	(5)	(6)
1	1st	9	(1-1),(2-1),(3-1),(4-1),(5-1),(6-1),(7-1), (8-1),(9-1)	(1-1)	(1-1)
	2nd	16	(1-2),(2-2),(3-2),(4-2),(5-2),(6-2),(7-2), (8-2),(9-2),(10-2),(11-2),(12-2),(13-2), (14-2),(15-2),(16-2)	(1-2)	(1-2)
2	1st	20	(1-1),(2-1),(3-1),(4-1),(5-1),(6-1),(7-1), (8-1),(9-1),(10-1),(11-1),(12-1),(13-1), (14-1),(15-1),(16-1),(17-1),(18-1), (19-1),(20-1)	(1-1)	(1-1)
	2nd	9	(1-2),(2-2),(3-2),(4-2),(5-2),(6-2),(7-2), (8-2),(9-2)	(1-2)	(1-2)
	3rd	20	(1-3),(2-3),(3-3),(4-3),(5-3),(6-3),(7-3), (8-3),(9-3),(10-3),(11-3),(12-3),(13-3), (14-3),(15-3),(16-3),(17-3),(18-3), (19-3),(20-3)	(1-3)	(1-3)
	4th	20	(1-4),(2-4),(3-4),(4-4),(5-4),(6-4),(7-4), (8-4),(9-4),(10-4),(11-4),(12-4),(13-4), (14-4),(15-4),(16-4),(17-4),(18-4),(19-4), (20-4)	(1-4)	(1-4)
3	1st	20	(1-1),(2-1),(3-1),(4-1),(5-1),(6-1),(7-1), (8-1),(9-1),(10-1),(11-1),(12-1),(13-1), (14-1),(15-1),(16-1),(17-1),(18-1),(19-1), (20-1)	(1-1)	—
	2nd	20	(1-2),(2-2),(3-2),(4-2),(5-2),(6-2),(7-2), (8-2),(9-2),(10-2),(11-2),(12-2),(13-2), (14-2),(15-2),(16-2),(17-2),(18-2),(19-2), (20-2)	(1-2)	(2-2),(3-2), (4-2),(5-2), (6-2),(7-2), (8-2)
	3rd	20	(1-3),(2-3),(3-3),(4-3),(5-3),(6-3),(7-3), (8-3),(9-3),(10-3),(11-3),(12-3),(13-3), (14-3),(15-3),(16-3),(17-3),(18-3),(19-3), (20-3)	(1-3)	—
	4th	20	(1-4),(2-4),(3-4),(4-4),(5-4),(6-4),(7-4), (8-4),(9-4),(10-4),(11-4),(12-4),(13-4), (14-4),(15-4),(16-4),(17-4),(18-4),(19-4), (20-4)	(1-4)	—
	5th	20	(1-5),(2-5),(3-5),(4-5),(5-5),(6-5),(7-5), (8-5),(9-5),(10-5),(11-5),(12-5),(13-5), (14-5),(15-5),(16-5),(17-5),(18-5),(19-5), (20-5)	(1-5)	—
10	1st	8	(1-1),(2-1),(3-1),(4-1),(5-1),(6-1),(7-1),(8-1)	(1-1)	(1-1)
	2nd	20	(1-2),(2-2),(3-2),(4-2),(5-2),(6-2),(7-2),(8-2), (9-2),(10-2),(11-2),(12-2),(13-2),(14-2), (15-2),(16-2),(17-2),(18-2),(19-2),(20-2)	(1-2)	(1-2)

(Continued)

TABLE 3.15 (*Continued*)

Representation of MPP for an Example Production Order of Six Part Types

Part Types	SQ No.	No. of MPP	Representation of Process Plans According to Gene Structure	Genes in 1st Iteration	Genes in 2nd Iteration
(1)	(2)	(3)	(4)	(5)	(6)
	3rd	20	(1-3),(2-3),(3-3),(4-3),(5-3),(6-3),(7-3), (8-3),(9-3),(10-3),(11-3),(12-3),(13-3), (14-3),(15-3),(16-3),(17-3),(18-3),(19-3), (20-3)	(1-3)	(1-3)
11	1st	6	(1-1),(2-1),(3-1),(4-1),(5-1),(6-1)	(1-1)	(1-1)
	2nd	15	(1-2),(2-2),(3-2),(4-2),(5-2),(6-2),(7-2), (8-2),(9-2),(10-2),(11-2),(12-2),(13-2), (14-2),(15-2)	(1-2)	(1-2)
	3rd	20	(1-3),(2-3),(3-3),(4-3),(5-3),(6-3),(7-3), (8-3),(9-3),(10-3),(11-3),(12-3),(13-3), (14-3),(15-3),(16-3),(17-3),(18-3),(19-3), (20-3)	(1-3)	(1-3)
12	1st	15	(1-1),(2-1),(3-1),(4-1),(5-1),(6-1),(7-1), (8-1),(9-1),(10-1),(11-1),(12-1),(13-1), (14-1),(15-1)	(1-1)	(1-1)
	2nd	8	(1-2),(2-2),(3-2),(4-2),(5-2),(6-2),(7-2), (8-2)	(1-2)	(1-2)
	3rd	15	(1-3),(2-3),(3-3),(4-3),(5-3),(6-3),(7-3), (8-3),(9-3),(10-3),(11-3),(12-3),(13-3), (14-3),(15-3)	(1-3)	(1-3)
	4th	7	(1-4),(2-4),(3-4),(4-4),(5-4),(6-4),(7-4)	(1-4)	(1-4)
	5th	20	(1-5),(2-5),(3-5),(4-5),(5-5),(6-5),(7-5), (8-5),(9-5),(10-5),(11-5),(12-5),(13-5), (14-5),(15-5),(16-5),(17-5),(18-5),(19-5), (20-5)	(1-5)	(1-5)

problems of maximisation characteristics of objective function (Deb, 2006). Although, the present work utilises the minimisation objective to reduce makespan value, i.e., $f(x)$. Therefore, the problem has been converted into maximisation through the transformation rule given by Deb (2006). Thus, $F(x) = 1/[1 + f(x)]$, where $f(x)$ is the value of makespan and $F(x)$ is the fitness function.

In the present work, simulation is used to calculate makespan value $[f(x)]$ for each process plan combination of various part types in a production order. Mathematical functions are less preferred as compared to simulation runs for a real-world system. This is because simulation performs close to an exact situation of a system. In the present work, ProModel® simulation software (www.promodel.com/) is used for the simulation and computation of makespan value. It is important to mention that during fitness evaluation, in case varying values of genes with similar chromosomes produce fitness values equivalent to the best chromosomes of earlier population, it is taken as an alternative optimum solution of the previous population.

3.6.4 Reproduction

This operator is the central part of the GA (Mitchell, 2002). In fact, the two main properties of GA viz., the offspring's production capability and the survival competence are represented by the reproduction operation of GA. During this process, the conception of fresh iteration begins if the optimisation norm deviates from a predetermined level. The fitness value of chromosomes is used to choose the parent individuals in order to create a new chromosome. A predetermined probability of mutation and crossover methods is utilised to emerge better offspring individuals. Thus, the value of fitness of each reproduced chromosome is calculated again. Finally, the parent population is replaced with the offspring, producing a fresh iteration (generation). In general, the GA maintained the fixed size of the population throughout the evolution process. This iterative cycle is continuing up until the predetermined optimisation norms are attained (Goldberg, 1989; Mitchell, 2002). Subsequent sections discuss the reproduction operators of GA such as selection, crossover, mutation, elitism, etc., used in the present work.

3.6.4.1 Selection Operation

This operator works for finding the individuals who undertake mutation and crossover operations according to the fitness values of parent individuals. The selection process attempts to explore the search in a better area of solutions using comparison methods corresponding to the fitness value of strings in order to select the good solutions, repeatedly. In literature, different types of selection procedures have been presented in order to solve job shop scheduling, like *Roulette Wheel Selection* (RWS), *Tournament Selection* (TS), and *Linear Ranking Selection* (LRS). This work merges LRS with *Stochastic Universal Sampling* (SUS). Mitchell (2002) recommended *LRS* with SUS as a good strategy in order to evade early merging of GA.

In LRS operation, the strings are ordered as per the fitness values. The expected value of an individual depends on its rank rather than on its absolute fitness (Baker, 1985). Baker (1985) proposed that the expected values of ordered population can be altered provided it is greater than or equal to zero. At each generation the individuals in the population are ranked in increasing order of fitness value and then assigned expected values according to the equation given below (Goldberg and Deb, 1991; Mitchell, 2002; Chakraborty, 2010).

$$Expected\ value\ (i) = Min + \left\{(Max - Min)\big(Rank\ (i) - 1\big)\right\}\big/\big(pop\_size - 1\big) \qquad (3.1)$$

where *Min* is the expected value of an individual with rank one, *pop_size* is the size of population, the constraint $Max \geq 0$, $1 \leq Max \leq 2$ for selection bias [i.e., higher values of the selection pressure (*Max*) cause the system to focus more on selecting only the better individuals] and *Min* = 2 – *Max*. In the present work, the selection pressure is taken as 1.6 as recommended by Baker (1985).

Once the expected values are assigned, SUS method is applied to sample the population (i.e., choose parents). It provides zero bias and minimum spread (Baker, 1985). Equally spaced random numbers are generated, as many as there are individuals to be selected. The random number is generated in the range "0" to "$1/n$" (Mitchell, 2002).

The working of selection method is explained for an example production order of six-part types as shown in Table 3.15. Column (1) of Table 3.16 presents an initial population of ten strings at *i*th generation (say) during the first iteration of integration methodology for an example production order. Column (3) presents fitness values of each string. The population is ranked in increasing order of their fitness values (column 4) and expected values are assigned to ranked population using equation number 1 (column 5). Now, SUS is applied on the population in order to select ten individuals for mating. Column (6) presents cumulative probability of selection. Column (7) presents that a single (first) random number has been generated between [0, $1/n$] and subsequent random numbers are generated by a unit increment to first random number. Now, in order to form a mating pool according to the SUS method, the string number 9th is selected 2 times and 6th is not selected (column 8). Thus, the mating population consists of strings number 7th, 10th, 4th, 3rd, 5th, 1st, 2nd, 9th (two times) and 8th.

3.6.4.2 Crossover

It is a process of recombining of two individuals and called as recombination or slice-exchange-merge operator. The aim of crossover operation is to recombine the parts of good substrings from parent strings in order to create a better offspring. It is based on the idea of breeding of two parents to produce a single child that has features from both parents, and thus may be better or worse than either parent as per fitness function (Goldberg, 1989). Various types of crossover operators such as partially matched crossover, order crossover, cycle crossover, single point crossover, two-point crossover, uniform crossover are used by researchers for job shop scheduling and IPPS. In the present study, two-point random crossover approach is applied on the individuals of mating pool. For a two-point crossover, two strings are selected randomly from the mating pool to make a pair. For each pair, the essentiality of carrying out a crossover is determined using the crossover probability (p_c = 0.8) as suggested by Mitchell (2002). It means that only $100 \times p_c$ percent strings in the population are used in the crossover operation and $100 \times (1 - p_c)$ percent of population remains as they are in the current population. Moreover, crossover sites are selected randomly from the first to the last position.

TABLE 3.16

Working of Selection Method for an Example Production Order

String No.	Initial Population (First Iteration of Integration Methodology)	Fitness Value {F(x)}	Order of String with Increasing Fitness Values	Expected Value Assigned to Ordered Strings	Cumulative Probability of Selection	Random Number	Actual Count in the Mating Pool (From SUS)
(1)	(2)	(3)	(4)	(5)	(6)	(7)	(8)
1st	(1-1) (1-2) (1-5) (1-2) (1-2) (1-3)	0.00281	7th	0.4000	0.4000	0.05328	1
2nd	(1-2) (1-1) (1-4) (1-3) (1-1) (1-4)	0.00367	6th	0.5333	0.9333	1.05328	0
3rd	(1-2) (1-2) (1-5) (1-2) (1-3) (1-5)	0.00173	10th	0.6666	1.5999	2.05328	1
4th	(1-1) (1-4) (1-1) (1-3) (1-1) (1-4)	0.00138	4th	0.7999	2.3999	3.05328	1
5th	(1-2) (1-2) (1-5) (1-2) (1-2) (1-3)	0.00230	3rd	0.9333	3.3331	4.05328	1
6th	(1-1) (1-3) (1-3) (1-1) (1-1) (1-2)	0.00103	5th	1.0666	4.3999	5.05328	1
7th	(1-2) (1-1) (1-5) (1-3) (1-3) (1-2)	0.00082	1st	1.1999	5.5999	6.05328	1
8th	**(1-1) (1-2) (1-4) (1-2) (1-2) (1-4)**[a]	**0.00533**	2nd	1.3333	6.9326	7.05328	1
9th	(1-1) (1-2) (1-2) (1-2) (1-2) (1-5)	0.00490	9th	1.4666	8.3992	8.05328	2
10th	(1-2) (1-4) (1-4) (1-1) (1-1) (1-2)	0.00120	8th	1.6000	9.9999	9.05328	1

[a] Indicates selection of string due to elitism.

TABLE 3.17

Working of Crossover Operation for an Example Production Order

String No.	Population After Selection	Sequence of Random Numbers (r)	Mating Pool (Parent Strings)	Crossing Sites (r_site)	After Crossover (Child Strings)
(1)	(2)	(3)	(4)	(5)	(6)
1st	(1-1) (1-2) (1-5) (1-2) (1-2) (1-3)	0.1310	(1-1) (1-2) (1-5) (1-2) (1-2) (1-3)	2-5	(1-1) **(1-4) (1-1)** **(1-3) (1-1)** (1-3)
2nd	(1-1) (1-2) (1-2) (1-2) (1-2) (1-5)	0.9830	—	—	—
3rd	(1-2) (1-2) (1-5) (1-2) (1-3) (1-5)	0.8930	—	—	—
4th	(1-1) (1-4) (1-1) (1-3) (1-1) (1-4)	0.0132	(1-1) (1-4) (1-1) (1-3) (1-1) (1-4)	2-5	(1-1) **(1-2) (1-5)** **(1-2) (1-2)** (1-4)
5th	(1-2) (1-2) (1-5) (1-2) (1-2) (1-3)	0.6321	(1-2) (1-2) (1-5) (1-2) (1-2) (1-3)	3-5	(1-2) (1-2) **(1-3)** **(1-1) (1-1)** (1-3)
6th	(1-1) (1-3) (1-3) (1-1) (1-1) (1-2)	0.7832	(1-1) (1-3) (1-3) (1-1) (1-1) (1-2)	3-5	(1-1) (1-3) **(1-5)** **(1-2) (1-2)** (1-2)
7th	(1-2) (1-1) (1-5) (1-3) (1-3) (1-2)	0.8321	—	—	—
8th	(1-1) (1-2) (1-4) (1-2) (1-2) (1-4)	0.3210	(1-1) (1-2) (1-4) (1-2) (1-2) (1-4)	4-6	(1-1) (1-2) (1-4) **(1-1) (1-1) (1-2)**
9th	(1-1) (1-2) (1-2) (1-2) (1-2) (1-5)	0.9210	—	—	—
10th	(1-2) (1-4) (1-4) (1-1) (1-1) (1-2)	0.2936	(1-2) (1-4) (1-4) (1-1) (1-1) (1-2)	4-6	(1-2) (1-4) (1-4) **(1-2) (1-2) (1-4)**

Note: Boldface indicates genes under crossover site.

Table 3.17 presents the working of adopted two-point random crossover operator. Column (2) presents the population that is to undergo crossover. For each string in the population, a random number (r) in the range [0, 1] is generated (column 3). If, $r \leq 0.8$ the string is selected for crossover. Thus, string number 1, 4, 5, 6, 8, and 10 are selected for mating as presented in column (4). Here, the number of selected strings is even, so the pairing is easy. If the numbers of selected strings are odd, either one extra string is added, or one selected string is removed, and this choice is made randomly. Now, the selected strings mate randomly; for example, the first two (i.e., 1st and 4th stings), the next two (i.e., 5th and 6th stings), and so on are coupled together. For each of these pairs, a random integer number (r_site) is generated in the range of [1, 6] (as 6 is length or number of genes in the string) and r_site indicates position of the crossing points (column 5). The first pair of strings (1st and 4th strings) undergoes crossover with $r\_site = 2$–5. It means that a segment of 1st string cut from 2nd to 5th gene and replaced by the same segment of 4th string (column 6). Once, the crossover of selected pairs is performed, children strings are inserted in the population for next operation.

3.6.4.3 Mutation

The aims of mutation operation are to achieve local search around the current solution as well as to maintain diversity in the population. Therefore, it performed on a bit-by-bit basis for a small random perturbation of the solutions (Mitchell, 2002). It is a background operator that helps to avoid premature convergence. During mutation, the genes are copied from the current string to the new string and each gene of the population undergoes mutation with quite a small mutation probability (p_m). Various types of mutation operators such as point mutation, swap mutation, exchange mutation, random mutation are used by researchers for job shop scheduling and IPPS. The present work utilises a random-exchange type of mutation operator with mutation probability ($p_m = 0.2$) and it is applied on population including offspring produced after crossover. $100 \times p_m$ percent strings in the population undergo mutation operation and $100 \times (1 - p_m)$ percent of the population remains same as they are in the current population. Moreover, the mutation position (from first to last gene) is selected randomly twice and process plans at these sites are interchanged, keeping other genes unchanged.

Table 3.18 presents the working of the random-exchange type mutation used in present work. Column (2) presents the intermediate population that is to undergo mutation operation. Sequences of random numbers (r) are generated in the range of 0–1 in column (3). Here, $p_m = 0.2$ and it means if $r \leq 0.2$ then three strings (1st, 4th and 10th) are affected by the mutation operator (column 4). In order to find exchange positions, the gene numbers within string are selected randomly twice in the range of 1–6 for a string (column 5). Thus, 1st string exchanges 2nd and 6th genes. The muted genes are typed in boldface (column 6).

3.6.5 Repairing Scheme

After mutation, some illegal offspring may generate, i.e., one-part type may exceed the limit of given MPP in the individuals which undergoes mutation operation. Thus, a repairing strategy is required to resolve this illegitimacy of offspring. In the present work, after mutation, all individuals who undergo mutation are checked to ensure that there is no part type that exceeds the limit range of process plans. If any position/site of the string exceeds the limit of available process plans, it is replaced by any one of the process plan from the given range of process plans randomly.

For example, column (2) in Table 3.19 presents the strings that are muted. Column (3) shows the position of genes that are muted in a string that undergo repairing procedure. The string numbered 10th contained an illegitimacy at fourth gene that belongs to part type 10 as per availability of MPP presented in column (5) of Table 3.15. Part type 10 contains maximum three SQs, while after mutation it shows 4th SQ and thus it has become illegal. Hence, the value of the fourth gene can be replaced by (1-1), (1-2), and (1-3) randomly that belongs to part type 10. Column (4) presents the strings after repairing.

TABLE 3.18

Working of Mutation Operation for an Example Production Order

String No.	Population After Selection and Crossover Operation	Sequence of Random Numbers (r)	Mating Pool (Parent Strings)	Exchange Positions	After Mutation
(1)	(2)	(3)	(4)	(5)	(6)
1st	(1-1) (1-4) (1-1) (1-3) (1-1) (1-3)	0.1243	(1-1) (1-4) (1-1) (1-3) (1-1) (1-3)	2 and 6	(1-1) **(1-3)** (1-1) (1-3) (1-1) **(1-4)**
2nd	(1-1) (1-2) (1-2) (1-2) (1-2) (1-5)	0.8837	—	—	—
3rd	(1-2) (1-2) (1-5) (1-2) (1-3) (1-5)	0.8735	—	—	—
4th	(1-1) (1-2) (1-5) (1-2) (1-2) (1-4)	0.0233	(1-1) (1-2) (1-5) (1-2) (1-2) (1-4)	1 and 5	**(1-2)** (1-2) (1-5) (1-2) **(1-1)** (1-4)
5th	(1-2) (1-2) (1-3) (1-1) (1-1) (1-3)	0.7322	—	—	—
6th	(1-1) (1-3) (1-5) (1-2) (1-2) (1-2)	0.4836	—	—	—
7th	(1-2) (1-1) (1-5) (1-3) (1-3) (1-2)	0.7324	—	—	—
8th	(1-1) (1-2) (1-4) (1-1) (1-1) (1-2)	0.5213	—	—	—
9th	(1-1) (1-2) (1-2) (1-2) (1-2) (1-5)	0.5212	—	—	—
10th	(1-2) (1-4) (1-4) (1-2) (1-2) (1-4)	0.1934	(1-2) (1-4) (1-4) (1-2) (1-2) (1-4)	2 and 4	(1-2) **(1-2)** (1-4) **(1-4)** (1-2) (1-4)

Note: Boldface indicates muted genes.

TABLE 3.19

Working of Repairing Scheme for an Example Production Order

String No.	Muted Strings	Genes to Undergo Repairing	Repaired Strings
(1)	(2)	(3)	(4)
1st	(1-1) (1-3) (1-1) (1-3) (1-1) (1-4)	—	—
2nd	—	—	—
3rd	—	—	—
4th	(1-2) (1-2) (1-5) (1-2) (1-1) (1-4)	—	—
5th	—	—	—
6th	—	—	—
7th	—	—	—
8th	—	—	—
9th	—	—	—
10th	(1-2) (1-2) (1-4) **(1-4)** (1-2) (1-4)	**(1-4)**	(1-2) (1-2) (1-4) *(1-2)* (1-2) (1-4)

Note: Boldface indicates muted genes that undergo repairing and italic-bold represents repaired genes.

After the generation of new population, the fitness value of each chromosome is computed using ProModel® simulation software. The higher the fitness value, the better is the performance of chromosome (i.e., parent). Hence, parents with higher fitness values have more chances to survive.

3.6.6 Elitism

After generating offspring, the parent strings of the previous generation are completely replaced. Therefore, a population with size of *pop_size* generates an equal number of offspring, as the best individual of population may fail to reproduce offspring in the next generation. Thus, elitism strategy is used in order to force GA to retain some number of the best individuals at each generation. It was first introduced by De Jong (1975). The best individuals can be lost in two cases: (i) if, they are not selected to reproduce and/or (ii) they are destroyed by crossover or mutation (Mitchell, 2002). During each new generation, the fraction of best individuals from previous generation into next generation has been called the *Generation Gap* (De Jong, 1975; Morad and Zalzala, 1999; Mitchell, 2002). In the present work, elitism transfers a good individual from the previous population to the population of the next generation with the elitism rate (*e_rate*) of 0.9 and it means that 10% best population is carried into the next generation, as applied by Morad and Zalzala (1999).

For example, if, *pop_size* = 10, then the total number of best individuals from the previous generation to be carried into the next generation is equal to one. Table 3.16 presents the population of ten strings in which the string number eight yields the best fitness value (0.0533). Therefore, according to the elitism selection strategy, the string number eight is carried into the next generation.

3.6.7 Termination Criterion

After initialisation, the iterative cycles of selection, crossover, and mutation are terminated when certain conditions are met and after termination, the elite chromosome passes the best solution found so far. Generally, there are two types of criteria that are used to terminate GA. First, if the elite solution has not changed for a certain number of generations, and second, if the number of generations has reached a maximum number (Mitchell, 2002). The present work utilised a maximum number of generations (max_gen) as a termination criterion. It depends on the size of the problem and is equal to the number of machines (m) times the number of part types (n), i.e., [$m \times n$], and hence allows more time to run as the size of the problem increases.

3.6.8 Restart Scheme

There are two important issues in optimal solution search strategies, (i) exploiting the best solution, (ii) exploring the search space. GA can strike a balance between exploration and exploitation of the search space. During the reproduction phase, crossover and mutation operations control exploitation and exploration of search space, respectively (Mitchell, 2002). The premature convergence occurs when the population achieved a sufficiently low diversity and entrapped at around a local optimum. Several techniques have been proposed to control premature convergence such as pre-selection, crowding (Goldberg, 1989), and fitness sharing (Goldberg and Deb, 1991). In the present work, to overcome premature convergence in the population, a restart scheme based on the ideas of a similar scheme used by Ruiz and Maroto (2006) is used. Accordingly, at each generation the best fitness value is stored. If the best fitness value is not changed for more than a pre-specified number of generations (*best_count*), the restart phase commences to regenerate the population. In the present work, the value of counter (*best_count*) is set at 15, i.e., if the highest fitness value in the population does not change for more than 15 generations, restart phase activates (Phanden et al., 2013).

Table 3.20 illustrates the working of a restart scheme for an example production order. Let us assume *best_count = 15* and the best fitness value = 0.0732 is not changed, therefore the restart phase will be activated. Column (2)

TABLE 3.20

Working of Restart Scheme for an Example Production Order

String No.	Population at *i*th Generation	Fitness Value	Population in Descending Order of Fitness Value (Step-1)		New Population	Fitness Value
(1)	(2)	(3)	(4)		(5)	(6)
1st	(1-2) (1-3) (1-4) (1-2) (1-1) (1-3)	0.0732	1st		(1-1) (1-2) (1-4) (1-2) (1-1) (1-3)	0.0431
2nd	(1-2) (1-3) (1-4) (1-2) (1-1) (1-3)	0.0732	2nd		(1-2) (1-1) (1-4) (1-2) (1-3) (1-3)	0.0680
3rd	(1-2) (1-3) (1-4) (1-2) (1-1) (1-3)	0.0732	3rd	First half (50%) Step-3.1	(1-2) (1-3) (1-4) (1-3) (1-1) (1-2)	0.0629
4th	(1-2) (1-3) (1-4) (1-2) (1-1) (1-3)	0.0732	4th		(1-1) (1-3) (1-4) (1-2) (1-2) (1-3)	0.0489
5th	(1-1) (1-2) (1-2) (1-3) (1-2) (1-5)	0.0489	7th		(1-2) (1-4) (1-3) (1-2) (1-1) (1-3)	0.0589
6th	(1-2) (1-2) (1-4) (1-3) (1-1) (1-3)	0.0701	8th		(1-1) (1-2) (1-1) (1-3) (1-2) (1-4)	0.0397
7th	(1-2) (1-3) (1-4) (1-2) (1-1) (1-3)	0.0732	9th		(1-2) (1-1) (1-3) (1-2) (1-1) (1-3)	0.0603
8th	(1-2) (1-3) (1-4) (1-2) (1-1) (1-3)	0.0732	6th		(1-1) (1-3) (1-2) (1-3) (1-3) (1-5)	0.0456
9th	(1-2) (1-3) (1-4) (1-2) (1-1) (1-3)	0.0732	5th	Second half (50%) Step-3.2	(1-2) (1-1) (1-4) (1-1) (1-3) (1-5)	0.0421
10th	(1-1) (1-3) (1-4) (1-1) (1-1) (1-4)	0.0394	10th		(1-2) (1-1) (1-4) (1-2) (1-2) (1-1)	0.0528

presents a *pop_size* of ten that undergoes restart procedure. It is sorted in decreasing order of fitness value (column 4). Then, first 20% strings (i.e., out of ten sorted strings, two strings – string number 1 and 2 at first and second positions, respectively) are deleted from the population and the remaining 80% (i.e., out of ten sorted strings – string numbers 3, 4, 7, 8, 9, 6, 5, and 10) are regenerated. The first half (50% of the population) is generated through random exchange mutation of the skipped string numbers 1 and 2 in column (5). Also, the other half (i.e., remaining 50% of the population) is generated randomly. The newly generated strings should possess fitness value greater than the fitness value of worst individual (i.e., fitness value of string number 10 = 0.0394 of the ith generation. Column (6) presents the fitness value of the new population for ith generation. Phanden et al. (2013) provide the details of various GA operators and their type/values used in the present work.

3.7 Case Study

The formalised IPPS approach is implemented in a case study for makespan performance measures. This case study is taken from the testbed problems reported by Kim et al. (2003). The present work considers testbed problem number 1 having 6-part types (Kim et al., 2003). Production quantity of each part type is generated randomly between 10 and 50. For this case study, a job shop consisting of 15 machines and an incoming location to receive the production order to be processed is taken into consideration. A uniform transportation time of three units is assumed between inter-machines and incoming location.

This considers testbed problem 1 of Kim (2003), which comprises part type 1, 2, 3, 10, 11, and 12. The production order details are referred from Phanden et al. (2013). These details of production orders are input to PPSM. It ranks MPP of each part type following minimum TPT criterion and stores the MPP in *Ranked MPP Database*. The best process plan from each SQ for each part type is retrieved from the database. Table 3.21 presents the output of PPSM. As soon as the best process plans and production quantity of each part type are received, SM is invoked. It generates an optimal schedule following simulation-based GA approach as discussed in the previous section. Completion time of each part type of production order in original schedule is presented in Table 3.22 and makespan of the production order is 1,418 min. Figure 3.4 shows the convergence curve of GA for the best fitness values during original schedule generation. The best fitness value of original schedule is 0.000705 at the termination criterion. Table 3.23 presents the selected process plans of each part type of production order in original schedule. The information generated by SM is sent to SAM for further analysis. At this stage, a human scheduler performs analysis of the schedule as per SAM

TABLE 3.21

Output of Process Plans Selection Module

S. No.	Part Type	SQ No.	Best Process Plans $\{M_i\ (T_i)\}$	TOT	TTT	TPT
(1)	(2)	(3)	(4)	(5)	(6)	(7)
1	1	1st	14(10)-15(61)-12(43)-11(39)-8(47)	200	18	218
2		2nd	14(10)-15(18)-13(30)-12(25)-1(40)-8(47)	170	21	191
3	2	1st	5(10)-9(10)-1(33)-5(38)	91	15	106
4		2nd	5(10)-8(6)-4(40)-5(38)	94	15	109
5		3rd	6(25)-10(41)-14(24)-6(38)-10(37)	165	18	183
6		4th	6(25)-10(41)-2(10)-4(25)-7(39)-10(37)	177	21	198
7	3	1st	4(29)-2(19)-7(21)	69	12	81
8		2nd	4(29)-8(29)-6(9)-7(13)	80	15	95
9		3rd	5(39)-14(33)-2(11)-1(10)-12(16)	109	18	127
10		4th	5(39)-1(6)-5(30)-3(39)-2(29)-1(19)	162	21	183
11		5th	15(44)-10(20)-7(29)-3(20)	113	15	128
12	10	1st	4(33)-15(20)	53	9	62
13		2nd	13(44)-7(35)-9(31)-12(45)-14(17)	172	18	190
14		3rd	13(44)-12(76)-9(19)-12(45)-14(17)	181	18	199
15	11	1st	10(27)-1(40)	67	9	76
16		2nd	2(6)-13(36)	42	9	51
17		3rd	6(30)-5(73)-6(29)	132	12	144
18	12	1st	11(29)-15(44)-5(5)-12(41)-11(15)	134	18	152
19		2nd	11(29)-15(24)-2(42)-11(15)	110	15	125
20		3rd	3(22)-1(5)-9(39)-13(19)	85	15	100
21		4th	3(18)-6(5)-4(18)-7(39)-10(7)	87	18	105
22		5th	3(22)-7(10)-12(38)-13(19)	89	15	104

TABLE 3.22

Completion Time of Each Part Type

Part Type	Original Schedule	After 1st Modification (Selected Part Type-11)			After 2nd Modification (Selected Part Type-3)		
		(Process Plan Combination Number – 1)	(Process Plan Combination Number – 2)	(Process Plan Combination Number – 3)	(1st Solution)	(2nd Solution)	(3rd Solution)
1	1418	1082	1081	1090	1090	1081	1082
2	1201	1201	1201	1201	1201	1201	1201
3	606	571	606	571	1149	1149	1149
10	1274	1274	1274	1274	816	1152	816
11	468	542	542	816	1069	542	542
12	1104	1110	1104	1033	903	1023	1110

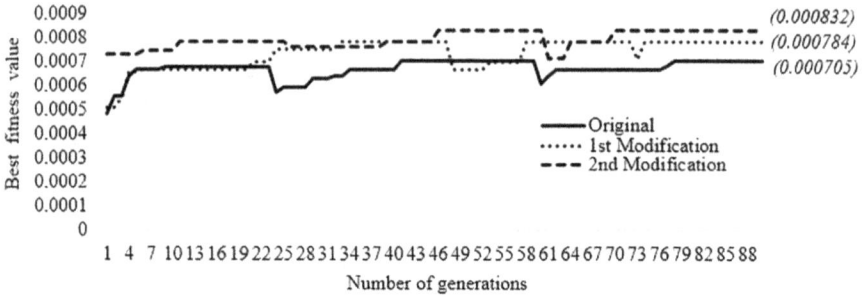

FIGURE 3.4
Convergence curve in schedule generation.

TABLE 3.23

Selected Process Plan of Each Part Type in Original Schedule

Part Type	SQ No.	Selected Process Plan $\{M_i\,(T_i)\}$	TOT	TTT	TPT
(1)	(2)	(3)	(4)	(5)	(6)
1	2	14(10)-15(18)-13(30)-12(25)-1(40)-8(47)	170	21	191
2	3	6(25)-10(41)-14(24)-6(38)-10(37)	165	18	183
3	1	4(29)-2(19)-7(21)	69	12	81
10	1	4(33)-15(20)	53	9	62
11	2	2(6)-13(36)	42	9	51
12	2	11(29)-15(24)-2(42)-11(15)	110	15	125

algorithm. It prepares the UTIMAC, UNUTIMAC, ALCPL, and OPERATION lists of current schedules. Here, ALCPL list is empty, as it is the first iteration of formalised IPPS approach. SAM reveals that *schedule improvement is possible*. Thus, a stage is set for process plan modification of a part type.

At this stage, PPMM is invoked and it searches a part type from the OPERATION list in decreasing order of priority in order to find the alternative process plan as per the PPMM algorithm for makespan. Here, part type 11 is selected for process plan modification. It follows second SQ and MPPs are retrieved from *Ranked MPP Database*. Columns 2 and 6 of Table 3.24 present available MPPs (15 no's) and fitness value as per scenario 1, respectively. It shows that seven alternative process plans are available as mentioned in S. No. 7, 9, and 11–15. All MPPs yield fitness value (0.000784) greater than the best fitness value (0.000705) of the previous schedule (original). Hence, as per the PPMM algorithm, all the MPPs of selected part type 11 are sent to SM and MPPs of other part types (1, 2, 3, 10, and 12) are the same as in the previous schedule.

Now, SM is invoked, and it generates the schedule with revised MPP of the selected part type 11. Table 3.22 shows completion time of each part type of production order after first modification for process plan combinations

TABLE 3.24

MPP of Selected Part Type-11 (Second Sequence of Operation)

S. No.	Multiple Process Plans $\{M_i (T_i)\}$	TOT	TTT	TPT	Fitness Value (Scenario 1)
(1)	(2)	(3)	(4)	(5)	(6)
1	2(6)-13(36)	42	9	51	—
2	**14**(7)-13(36)	43	9	52	—
3	7(8)-13(36)	44	9	53	—
4	2(6)-**9**(39)	45	9	54	—
5	2(6)-**5**(40)	46	9	55	—
6	**9**(10)-13(36)	46	9	55	—
7	**14**(7)-**9**(39)	46	9	55	*0.000784*
8	9(49)	49	6	55	—
9	7(8)-**9**(39)	47	9	56	*0.000784*
10	**11**(11)-13(36)	47	9	56	—
11	**14**(7)-**5**(40)	47	9	56	*0.000784*
12	7(8)-**5**(40)	48	9	57	*0.000784*
13	**9**(10)-**5**(40)	50	9	59	*0.000784*
14	**11**(11)-**9**(39)	50	9	59	*0.000784*
15	**11**(11)-**5**(40)	51	9	60	*0.000784*

Note: Boldface indicates under-utilised machines and bold-italics indicates that it is greater than value obtained in an earlier schedule; Scenario 1 indicates that one or more alternative process plans containing more than one under-utilised machine that yield fitness value greater than previous schedule.

1, 2, and 3. It reveals that makespan of the production order is reduced from 1,418 min in the original schedule to 1,274 min after modification of process plan of the part type 11.

Figure 3.4 presents the convergence curve of GA for the best fitness values after the first modification. The best fitness value is 0.000784 at the termination criterion. Table 3.25 presents the combinations of process plans of production order after first modification. SM identifies three sets of process plans for the production order and each combination yields the same makespan i.e., 1,274 min. Now, SAM is invoked, and it selects one process plan combination randomly. Here, process plan combination number (1) is selected by the scheduler to perform analysis. Further, SAM generates the UTIMAC, UNUTIMAC, ALCPL, and OPERATION lists for the selected schedule. At this stage, the ALCPL list contains part type 11 and the OPERATION list is prepared for remaining part types (1, 2, 3, 10, and 12). Hence, SAM indicates that *schedule improvement is possible* and relevant information is sent to PPMM.

PPMM searches a part type for process plans modification from the OPERATION list. It selects part type 3 and column (2) of Table 3.26 presents its MPP that are retrieved from *Ranked MPP Database*. Column (6) presents

TABLE 3.25

Selected Process Plan Combinations after First Modification (Selected Part Type-11)

PPC No.	Part Type	SQ No.	Selected Process Plan $\{M_i\,(T_i)\}$	TOT	TTT	TPT
(1)	(2)	(3)	(4)	(5)	(6)	(7)
1	1	2	14(10)-15(18)-13(30)-12(25)-1(40)-8(47)	170	21	191
	2	3	6(25)-10(41)-14(24)-6(38)-10(37)	165	18	183
	3	1	4(29)-2(19)-7(21)	69	12	81
	10	1	4(33)-15(20)	53	9	62
	11	2	9(49)	49	6	55
	12	3	3(22)-1(5)-9(39)-13(19)	85	15	100
2	1	2	**14(10)-15(18)-13(30)-12(25)-1(40)-8(47)**	170	21	191
	2	3	**6(25)-10(41)-14(24)-6(38)-10(37)**	165	18	183
	3	1	**4(29)-2(19)-7(21)**	69	12	81
	10	1	**4(33)-15(20)**	53	9	62
	11	2	**9(49)**	49	6	55
	12	2	**11(29)-15(24)-2(42)-11(15)**	110	15	125
3	1	2	14(10)-15(18)-13(30)-12(25)-1(40)-8(47)	170	21	191
	2	3	6(25)-10(41)-14(24)-6(38)-10(37)	165	18	183
	3	1	4(29)-2(19)-7(21)	69	12	81
	10	1	4(33)-15(20)	53	9	62
	11	2	14(7)-9(39)	46	9	55
	12	3	3(22)-1(5)-9(39)-13(19)	85	15	100

Note: PPC No., Process Plans Combination Number; Bold indicates that process plans combination followed by SAM.

TABLE 3.26

MPP of Selected Part Type-3 (First Sequence of Operation)

S. No.	Multiple Process Plans $\{M_i\,(T_i)\}$	TOT	TTT	TPT	Fitness Value (Scenario 1)
(1)	(2)	(3)	(4)	(5)	(6)
1	4(29)-2(19)-7(21)	69	12	81	—
2	4(29)-2(19)-**13(5)**-2(18)	71	15	86	—
3	4(51)-13(5)-2(18)	74	12	86	—
4	4(29)-**3(24)**-7(21)	74	12	86	0.000784
5	4(29)-2(19)-**13(5)-3(20)**	73	15	88	0.000784
6	4(29)-2(19)-10(12)-7(13)	73	15	88	—
7	**7(36)**-2(19)-7(21)	76	12	88	0.000689
8	4(51)-10(12)-7(13)	76	12	88	—
9	4(29)-2(19)-7(8)-2(18)	74	15	89	—
10	4(29)-2(19)-1(13)-**7(13)**	74	15	89	—
11	**11(34)-4(22)**-7(21)	77	12	89	0.000716
12	4(51)-1(13)-7(13)	77	12	89	—
13	4(29)-2(19)-**12(9)**-2(18)	75	15	90	—

(Continued)

TABLE 3.26 (*Continued*)

MPP of Selected Part Type-3 (First Sequence of Operation)

S. No.	Multiple Process Plans $\{M_i(T_i)\}$	TOT	TTT	TPT	Fitness Value (Scenario 1)
(1)	(2)	(3)	(4)	(5)	(6)
14	4(51)-**12**(9)-2(18)	78	12	90	—
15	4(29)-**3**(24)-**13**(5)-2(18)	76	15	91	0.000635
16	**11**(34)-2(19)-**13**(5)-2(18)	76	15	91	0.000689
17	4(29)-2(19)-7(8)-**3**(20)	76	15	91	0.000674
18	7(36)-4(22)-7(21)	79	12	91	*0.000832*
19	**11**(34)-**3**(24)-7(21)	79	12	91	0.000689
20	4(29)-2(19)-**12**(9)-**3**(20)	77	15	92	0.000689

Note: Boldface indicates under-utilised machines and bold-italics indicates that it is greater than value obtained in an earlier schedule; Scenario 1 indicates that one or more alternative process plans to contain more than one under-utilised machine that yields fitness value greater than previous schedule.

the fitness values with alternative process plans of selected part type as per scenario (1). There are ten alternative process plans available that are mentioned in S. No. 4, 5, 7 11, and 15–20, respectively. One alternative process plan shown in S. No. 18 [*7(36)-4(22)-7(21)*] yields a fitness value (0.000832) greater than the best fitness value (0.000784) of previous schedule. Thus, all MPP of selected part type 3 are passed to SM, keeping process plans of other part types same as in previous schedule (first modification).

After the second modification, the makespan of the production order is further reduced to 1,201 min from 1,274 min that was obtained after the first modification. SM provides three schedules and process plan combinations that yield the same makespan (1,201 min) and the completion time of each part type is shown in Table 3.22 for 1st, 2nd, and 3rd schedule. Figure 3.4 illustrates the convergence curve of GA for the best fitness value during revised schedule generation and Table 3.27 presents combinations of process plans of production order after second modification.

Now, SAM is invoked, and it randomly selects process plan combination number 1 for the analysis and prepares the UTIMAC, UNUTIMAC, ALCPL, as well as OPERATION lists. It reveals that schedule improvement is possible as the OPERATION list is not empty.

At this stage, PPMM is invoked, and it searches for the alternative process plans of a part type. Here, PPMM algorithm does not select a part type for the process plan modification as the available MPP of part types as per the OPERATION list yield fitness value lower than the best fitness value of current schedule. Thus, further improvement in makespan is not possible. Figure 3.5 shows makespan of a production order for each iteration of process plan modifications. Makespan of the production order is reduced by 10.15% in first modification and by 5.72% in the second modification and total improvement is 15.30%. Thus, the formalised integrated approach is

TABLE 3.27

Selected Process Plan Combinations After Second Modification (Selected Part Type-3)

PPC No.	Part Type	SQ No.	Selected Process Plans $\{M_i\,(T_i)\}$	TOT	TTT	TPT
(1)	(2)	(3)	(4)	(5)	(6)	(7)
1 (First Solution)	1	2	**14(10)-15(18)-13(30)-12(25)-1(40)-8(47)**	170	21	191
	2	3	**6(25)-10(41)-14(24)-6(38)-10(37)**	165	18	183
	3	1	**11(34)-4(22)-7(21)**	77	12	89
	10	1	**4(33)-15(20)**	53	9	62
	11	2	**11(11)-5(40)**	51	9	60
	12	3	**3(22)-1(5)-9(39)-13(19)**	85	15	100
2 (Second Solution)	1	2	14(10)-15(18)-13(30)-12(25)-1(40)-8(47)	170	21	191
	2	3	6(25)-10(41)-14(24)-6(38)-10(37)	165	18	183
	3	1	7(36)-4(22)-7(21)	79	12	91
	10	1	4(33)-15(20)	53	9	62
	11	2	9(49)	49	6	55
	12	3	11(29)-15(24)-2(42)-11(15)	110	15	125
3 (Third Solution)	1	2	14(10)-15(18)-13(30)-12(25)-1(40)-8(47)	170	21	191
	2	3	6(25)-10(41)-14(24)-6(38)-10(37)	165	18	183
	3	1	7(36)-4(22)-7(21)	79	12	91
	10	1	4(33)-15(20)	53	9	62
	11	2	9(49)	49	6	55
	12	3	3(22)-1(5)-9(39)-13(19)	85	15	100

Note: PPC No., Process Plans Combination Number; bold indicates that process plans combination followed by SAM.

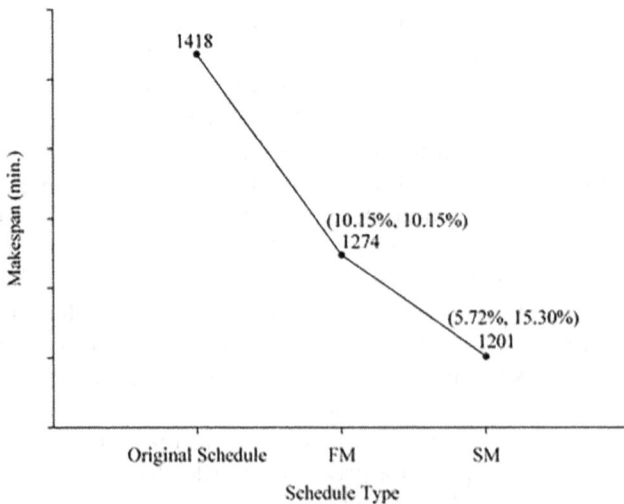

FIGURE 3.5
Makespan of production order.

effective in reducing the makespan of a production order for the present case. This case study also reveals that there are more than one process plan combinations of various part types of production orders that yield the same optimal makespan.

Legends: FM- schedule after first modification, SM- schedule after second modification (values in bracket indicate percentage improvement from preceding and original schedule, respectively).

3.8 Comparison of Formalised IPPS Methodology With Conventional Manufacturing Methodology

The performance of the formalised integration approach is compared with the hierarchical approach to assess the effectiveness of formalised integration approach over hierarchical approach. Traditionally, in hierarchical approach, process planning is first solved and then scheduling is attempted. The present work considers three scenarios in hierarchical approaches. In the first scenario, a process plan for each part type is selected based on the minimum TPT and it is represented as SPP (1). In the second scenario, a single process plan for each part type is selected based on the minimum number of setups and it is represented as SPP (2). The third scenario selects a process plan for each part type randomly and it is represented as SPP (3). All parameters and their values, assumptions, as well as genetic operators used in the hierarchical approach are the same as those used in the integration approach. Figure 3.6 illustrates the performance of formalised integration approach with hierarchical approach for case study. It clearly reveals that formalised integration methodology is more effective compared to hierarchical approach from a makespan viewpoint. The percentage improvement of formalised IPPS methodology over the best hierarchical methodology is 47.04%, 40.30%, and 47.60% from a single process plan selected with minimum TPT,

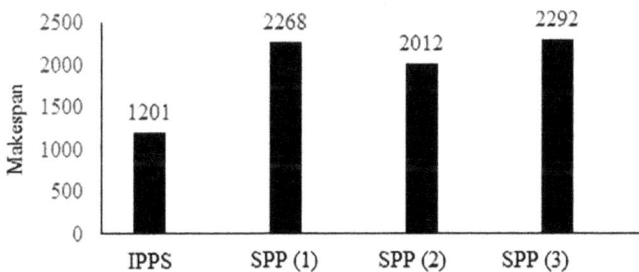

FIGURE 3.6
Performance of formalised integration approach with hierarchical approach.

minimum number of operation setups, and randomly selected criterion, respectively. The percentage improvement is computed as:

$$\text{Percentage improvement} = \frac{\left(\begin{array}{c}\text{Makespan in hierarchical}\\\text{approach}\end{array}\right) - \left(\begin{array}{c}\text{Makespan in formalized}\\\text{IPPS approach}\end{array}\right)}{(\text{Makespan in hierarchical approach})} \times 100$$

Thus, it can be safely concluded that formalised IPPS methodology is better than conventional hierarchical methodology for makespan.

Legends: SPP (1) – Single Process Plan selected with minimum TPT criterion; SPP (2) – Single Process Plan selected with minimum number of operation setups criterion; SPP (3) – Randomly selected Single Process Plan criterion.

3.9 Epilogue

This chapter presented an IPPS using simulation-based GA for makespan performance measure. Various modules such as PPSM, SM, SAM, and PPMM were presented with a suitable example. Also, the steps of heuristics for makespan measure have been explained in detail. Finally, the methodology has been illustrated in a case study. It has been concluded that the proposed IPPS outperformed conventional manufacturing methodology.

References

Baker, J. E., 1985. Adaptive selection methods for Genetic algorithms. Proceedings of the 1st International Conference on Genetic Algorithms and their Applications, pp. 101–111.

Chakraborty, R C., 2010. Genetic algorithm and modeling. *Soft Computing Course Lecture*, 37–40.

Chryssolouris, G., 2005. *Manufacturing Systems – Theory and Practice*. 2nd edition, Springer-Verlag, New York.

De Jong, K. A., 1975. An analysis of the behavior of a class of genetic adaptive systems. Doctoral dissertation. Dissertation Abstracts International, University of Michigan, 36(10), 5140B, No. 76-9381.

Deb, K., 2006. *Optimization for Engineering Design, Algorithm and Examples*. Prentice-Hall of India, New Delhi.

Goldberg, D. E. and Deb, K., 1991. *A Comparative Analysis of Selection Schemes Used in Genetic Algorithms. Foundations of Genetic Algorithms*, Morgan Kaufmann Publishers, San Mateo, California, USA, pp. 69–93.

Goldberg, D. E., 1989. *Genetic Algorithms in Search, Optimization, and Machine Learning*. Addison Wesley, Reading, MA.

Jain, A., Jain, P. and Singh, I., 2006. An integrated scheme for process planning and scheduling in FMS. *International Journal of Advanced Manufacturing Technology*, 30, 1111–1118.

Kim, Y., et al., 2003. A symbiotic evolutionary algorithm for the integration of process planning and job shop scheduling. *Computers and Operations Research*, 30, 1151–1171.

Kim, Y.K., 2003. A set of data for the integration of process planning and job shop scheduling [online]. Available from: http://syslab.chonnam.ac.kr/links/data-pp&s.doc [Assessed 7 November 2011].

Mitchell, M., 2002. *An Introduction to Genetic Algorithms*. Prentice-Hall, New Delhi, India.

Morad, N. and Zalzala, A., 1999. Genetic algorithms in integrated process planning and scheduling. *Journal of Intelligent Manufacturing*, 10, 169–179.

Phanden, R. K., Jain, A. and Verma, R., 2013. An approach for integration of process planning and scheduling. *International Journal of Computer Integrated Manufacturing*, 26(4), 284–302.

Ruiz, R. and Maroto, C., 2006. A genetic algorithm for hybrid flowshops with sequence dependent setup times and machine eligibility. *European Journal of Operational Research*, 169, 781–800.

4

An Approach to Integrated Process
Planning and Scheduling Based on
Variable Neighbourhood Search

Oleh Sobeyko and Lars Mönch

FernUniversität in Hagen

CONTENTS

4.1   Introduction 95
4.2   Problem Setting 96
4.3   Discussion of Previous Work 98
4.4   VNS-Based Approach 99
      4.4.1   Overall Setting 99
      4.4.2   Algorithmic Details 101
      4.4.3   Reference Heuristic 105
4.5   Simulation-Based Performance Assessment of the IPPS Approach 105
      4.5.1   Simulation Infrastructure 105
      4.5.2   Rolling Horizon Scheme 107
4.6   Simulation Experiments 108
      4.6.1   Design of Experiments 108
      4.6.2   Parameter Setting 111
      4.6.3   Simulation Results 111
4.7   Conclusion and Future Research Directions 112
References 113

4.1 Introduction

Process planning activities select the operations that must be performed to produce a certain product (Phanden et al. 2011). Moreover, precedence constraints among different operations and the routing of operations are determined by process planning. Scheduling is an activity that assigns the machines and secondary resources to the operations that are found by process planning, aiming to optimise specific performance measures. This results in a set of start times for operations on the machines, i.e., in a schedule.

Pure scheduling algorithms for flexible job shops assume that the route information for the different products is known. Only a single route is possible for each product. In real-world settings, however, this assumption is often not true. Apart from this, even alternative bills of materials are possible for certain products (Ivens & Lambrecht 1996). In addition, a set of possible routes determined by process planning might exist for each subpart of the bill of materials. This setting can be found, for instance, in the food industry (Sobeyko & Mönch 2017). Process planning is often performed before the scheduling activities take place. The availability of the resources and their finite capacity is not taken into account by process planning. As a consequence, process plans are often modified. This can be avoided when scheduling and process planning are carried out in an integrated manner since the scheduling execution is then anticipated when the process planning decisions are made.

In the present chapter, we consider an integrated process planning and scheduling (IPPS) approach for flexible job shops of large size with total weighted tardiness (TWT) objective. Alternative bills of materials and the corresponding routes are assumed for each product. After a bill of materials and corresponding routes are selected for each product, a scheduling problem remains to be solved. A variable neighbourhood search (VNS)-based scheme is applied to choose bills of materials and routes, while the iterative local search (LS) procedure proposed by Sobeyko and Mönch (2016) is taken to solve instances of a flexible job shop scheduling problem. While the outcome of extensive computational tests using randomly generated problem instances is reported in (Sobeyko & Mönch 2017), in the chapter at hand, we investigate the performance of the proposed IPPS scheme when it is performed by a rolling horizon approach. This requires simulation to represent the execution of the schedules in the manufacturing system. We know from the literature that assessing the performance of deterministic planning and scheduling approaches is possible in a more realistic way when they are applied in a rolling horizon manner (Mönch 2007, Ponsignon & Mönch 2014). IPPS approaches are only rarely assessed based on a rolling horizon approach.

The rest of the chapter is structured as follows. The next section deals with describing the problem at hand. Previous work is briefly surveyed in Section 4.3. The VNS scheme that is proposed is described in Section 4.4. The infrastructure for performance assessment activities based on simulation is discussed in Section 4.5. Results of simulation experiments are presented in Section 4.6. Conclusions and directions for future research are discussed in Section 4.7.

4.2 Problem Setting

Flexible job shops including m machine groups $M := \{M_1, \ldots, M_m\}$ are studied. Unrelated parallel machines, i.e., machines which offer the same functionality form a machine group. The operations that can be performed on the machines of M_i are $O(M_i)$. We assume that a unique M_i belongs to each operation, i.e., it holds $O(M_i) \cap O(M_j) = \emptyset$ for all $i \neq j$. Moreover, it is assumed that no preemption is possible when operations are executed on a machine.

Different products $P_1,\ldots,P_r$ are considered in this research. A bill of materials is given by a set of subparts. It describes the technological precedence constraints among subparts. Only bills of materials with a convergent structure are considered in this research. Alternative bills of materials B_{iz} are assigned to product P_i. This is a reasonable assumption since a certain subpart might be produced by different suppliers. A subpart can have alternative routes. A sequence of operations that must be processed consecutively to manufacture a subpart constitutes a route. Operations have to be processed on a machine of the machine group that belongs to the operation. We assume in this research that the alternative routes for a subpart are permutations of the same operations.

A selection of a bill of materials for product P_i and of routes for all subparts of this bill of materials forms a process plan for this product. We call this process plan PP_i. A process plan for products $P_1,\ldots,P_r$ is given by $PP = \{PP_1,\ldots,PP_r\}$. We will indicate this dependency from a given set of products if necessary, otherwise we will suppress it.

The s-th order belonging to product P_i is O_{is}. This order has a ready time r_{is}, a due date d_{is}, and a weight w_{is}. Moreover, C_{is} is the completion time of this order. Its tardiness is given by

$$T_{is} := \max(C_{is} - d_{is}, 0) \tag{4.1}$$

The performance measure TWT that has to be minimised is

$$TWT := \sum_{i,s} w_{is} T_{is} \tag{4.2}$$

The following three decisions must be made when an IPPS instance is solved:

1. A bill of materials has to be chosen for each product.
2. A route has to be selected for each subpart of the chosen bills of materials.
3. Operations resulting from the steps 1 and 2 decisions must be scheduled.

Using the standard $\alpha|\beta|\gamma$ representation scheme from deterministic scheduling, the IPPS problem can be formulated as

$$FJ|r_{is}, BOM, alter|TWT \tag{4.3}$$

where *FJ* refers to a flexible job shop in the α field and *BOM* and *alter* indicate the alternative bills of materials and routes in the β field. The IPPS problem (4.3) is NP-hard due to the fact that the NP-hard single-machine scheduling problem $1 \| TWT$ (Lawler 1977) is a special case of this problem. A mixed integer linear programming formulation for problem (4.3) is developed by Sobeyko and Mönch (2017). However, since problem (4.3) is NP-hard, optimal solutions can only be computed for problem instances with up to three orders within a reasonable maximum computing time (see Sobeyko & Mönch 2017).

Hence, efficient heuristics have to be used to tackle problem instances of medium and large size using a reasonable amount of time. Simple reference heuristics and a more advanced VNS-based approach are designed and computationally assessed by Sobeyko and Mönch (2017) using randomly generated problem instances.

However, in real-world environments, the IPPS approach will be applied in a rolling horizon manner respecting the fact that feedback from the dynamic and stochastic shop floor is required. In the present chapter, we aim to evaluate the performance of these heuristics when they are executed in a rolling horizon environment. It is well known from the literature on rolling horizon approaches (cf. Ponsignon & Mönch 2014 or Ziarnetzky et al. 2015) that the advantage of sophisticated heuristics for planning and scheduling does not necessarily carry over from a static and deterministic environment to a dynamic and stochastic environment. The magnitude of improvement of the more sophisticated approaches compared to simple approaches with less optimisation abilities is often smaller than in a static and deterministic setting.

4.3 Discussion of Previous Work

We discuss only related research for scheduling of complex job shops and IPPS approaches with respect to applying them in a rolling horizon environment. A detailed discussion of various IPPS approaches from the literature is provided in Chapter 2 of this volume.

Complex job shops are flexible job shops with additional process conditions such as reentrant flows, sequence-dependent setup times, and batch processing. Batching refers to forming groups of jobs to simultaneously process them on the same machine. These process conditions are motivated by semiconductor manufacturing (Mönch et al. 2013). Rolling horizon approaches for scheduling approaches using the shifting bottleneck heuristic (SBH) (cf. Adams et al. 1988 and Ovacik & Uzsoy 1997) are investigated (cf. Barua et al. 2005, Mönch & Drießel 2005, Mönch et al. 2007, Pfund et al. 2008, Upasani & Uzsoy 2008, Mönch & Zimmermann 2011, Driessel & Mönch 2012). The simulation infrastructure proposed by Mönch et al. (2003) is used in most of these papers. The backbone of the infrastructure is a data layer that is between the simulated shop floor, the base system, and the corresponding scheduling algorithm. The rolling horizon setting is crucial to demonstrate that deterministic global scheduling approaches are able to outperform myopic dispatching approaches in a dynamic and stochastic environment when they are correctly parameterised. However, only a single route exists per product in these papers. Despite this fact it is possible to reuse the simulation infrastructure by extending them to the IPPS setting.

Next, we discuss simulation applications for solving IPPS problems. Lee and Kim (2001) propose a genetic algorithm (GA) based on simulation for IPPS problems with makespan C_{max} and maximum lateness L_{max} performance measures. Discrete-event simulation is applied to compute performance measure values for the selected process plans by using dispatching rules for scheduling. An agent-based negotiation approach for IPPS is proposed by Wong et al. (2006). A job shop is assumed. The job and machine agents negotiate to select a process plan and to compute schedules. The negotiation process is assessed by discrete-event simulation. The mean flow time and C_{max} are considered as performance measures. A GA for tackling an IPPS problem is proposed by Phanden et al. (2013). The mean tardiness and C_{max} are considered as performance measures. Simulation is used to calculate the fitness value that belongs to the chromosomes. A GA for a remanufacturing IPPS problem is proposed by Zhang et al. (2015). The chromosomes are evaluated using Monte Carlo simulation since the number of jobs with specific quality features is uncertain.

Although simulation is used in specific IPPS approaches, we conclude that no work for IPPS exists where a rolling horizon scheme is used to assess the performance of the IPPS approach.

4.4 VNS-Based Approach

4.4.1 Overall Setting

According to Brandimarte and Calderini (1995), the IPPS problem (4.3) can be decomposed into the following subproblems:

1. Determination of process plans by choosing bills of materials and routes per product
2. Scheduling of the operations belonging to the subparts of the selected process plans

Taking into account the described decomposition, a process planner works together with a scheduler. The scheduler interacts with the shop floor. The process planner is responsible for the following tasks:

- It analyses feedback from the scheduler.
- It selects bills of materials.
- It selects routes.
- It sends a process plan to the scheduler.
- It terminates the process planning process.
- It informs the scheduler about terminating the process planning process.

The scheduler carries out the following tasks:

- It determines schedules based on the proposed process plan, i.e., a set of bill of materials and routes from the process planner.
- It calculates the performance measure value of the schedule.
- It sends the result of assessing the current process plan to the process planner.
- It sends the final schedule to the shop floor after the process planning process is terminated.

The decomposition scheme with the process planner and scheduler is depicted in Figure 4.1.

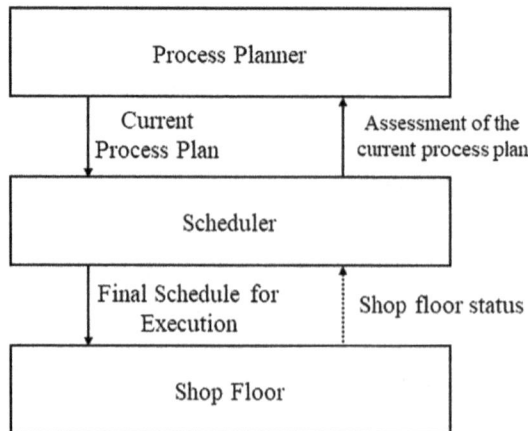

FIGURE 4.1
Decomposition scheme for the IPPS task.

The decomposition procedure works iteratively. An initial process plan is first selected. The scheduler determines the quality of this process plan by computing the *TWT* value of the schedule. The process planner iteratively determines new process plans based on the obtained performance measure value. The scheduler again evaluates these process plans. The proposed decomposition has the advantage that sophisticated algorithms can be used for the process planning and scheduling level.

4.4.2 Algorithmic Details

The VNS heuristic for the process planning level is described first. The meta-heuristic VNS extends a basic LS procedure to enable it to avoid local optima. When the LS scheme gets stuck in a non-global optimum, a randomly chosen neighbour of the incumbent solution is used to restart it. This is called shaking. Several neighbourhood structures are used within shaking (Hansen & Mladenovic 2001, 2003).

Different VNS variants exist. Here, we apply skewed reduced VNS (SRVNS). Reduced VNS avoids LS after the shaking step to reduce the computational burden that is a result of solving many flexible job shop scheduling problem instances. Skewed VNS is a VNS variant that allows for accepting moves that do not necessarily improve a solution. The general SRVNS algorithm for a combinatorial optimisation problem with performance measure f to be minimised can be described as follows:

SRVNS

1. Initialisation: Choose different neighbourhood structures N_k, $k = 1, \ldots, k_{\max}$, find an initial solution x and the corresponding $f(x)$ value. Initialise $x^* \leftarrow x$. A stopping condition and a parameter value λ are chosen. Initialise $k \leftarrow 1$.

2. Repeat the steps below until the stopping condition is met:

3. Shaking: Randomly choose a neighbour $x' \in N_k(x)$.

4. Improvement or not: Update $x^* \leftarrow x'$ if the best solution x^* is outperformed by x'.

5. Move or not: If $f(x') - \lambda \rho(x^*, x) < f(x)$ then update the incumbent solution by $x \leftarrow x'$. Moreover, set $k \leftarrow 1$, otherwise set $k \leftarrow k \bmod k_{\max} + 1$. Go to Step 2.

In Step 5 of the SRVNS scheme, $\lambda > 0$ is a scaling parameter, whereas the function ρ measures the distance of solutions. This step allows for accepting non-improving moves in certain situations. The search space on the process planning level is given by $PP = \{PP_1, \ldots, PP_r\}$.

Next, we specify the applied parameterised neighbourhood structures to tailor the SRVNS scheme for problem (4.3). Therefore, we introduce the following additional notation. The alternative bills of materials for product P_i, $i = 1,\ldots,r$ are called B_{iz}. The bill of materials B_{iz} has $|B_{iz}|$ subparts. The l-th route of the j-th subpart of the bill of materials B_{iz} is denoted by R_{izjl}. The quantity $|R_{izjl}|$ refers to the number of operations that must be performed to manufacture R_{izjl}. The following neighbourhood structures are used:

- $N_1(k)$: Randomly choose products $P_{i_1},\ldots,P_{i_k}$. For each these products P_i with bill of materials B_{iz}, choose randomly a bill of materials B_{iz^*} where $z \neq z^*$ holds if this is possible. For the j-th subpart belonging to B_{iz^*}, select the same route as for the j-th subpart in B_{iz}. Choose a route randomly if this is not possible.

- $N_2(k)$: Randomly choose products $P_{i_1},\ldots,P_{i_k}$. For each of these products P_i with bill of materials B_{iz}, choose randomly $s \sim DU[1,|B_{iz}|]$ subparts within this bill of materials where the notation $DU[a,b]$ stands for a discrete uniform distribution over $\{a,\ldots,b\}$. Choose randomly an alternative route per selected subpart.

- $N_3(k)$: A number of k different products $P_{i_1},\ldots,P_{i_k}$ are randomly chosen from $P_1,\ldots,P_r$ according to a discrete distribution with probabilities $p_i := \sum_s w_{is}T_{is}/\sum_{j,s} w_{js}T_{js}$ where the tardiness values are taken from the best solution already found. Note that the case $\sum_s w_{is}T_{is} = 0$ is not interesting since we already know an optimal solution in this situation. For each chosen P_i with bill of materials B_{iz}, choose an alternative bill of materials B_{iz^*}, $z^* \neq z$ if this is possible based on a discrete distribution with probabilities $p_{iz^*} := 1 - RPT_{iz^*}/\sum_z RPT_{iz}$. Here, RPT_{iz} is the estimated raw processing time (RPT) required to manufacture P_i based on the bill of materials B_{iz}. Moreover, the corresponding routes are chosen as carried out for the neighbourhood structure N_1.

We see that N_1 and N_3 explore the solution space by changing the bill of materials of a product. In contrast to this, only the routes are changed by N_2. Therefore, it is not guaranteed in general that a solution with minimum TWT value can be obtained using a finite number of moves based on these neighbourhood structures. To avoid this disadvantage, combined neighbourhood structures are designed for a prescribed k value, for instance, N_2N_1 or N_2N_3, i.e., a composition of mappings is used. The neighborhood structure N_3 is similar to N_1. Information on the scheduling process is taken into account when the bills of materials are chosen by N_3. If O_{is} causes a large value of the weighted tardiness then it is expected that changing the bill of materials and the corresponding routes is beneficial to get a better process plan. Moreover, bills of materials are ranked based on their RPT. It is expected that a bill of materials with a fairly small RPT value will lead to small waiting times for the operations. In addition the combined neighbourhood structures N_2N_1 or N_2N_3, we consider the combined neighbourhood structures N_3N_1, N_1N_3, $N_2N_3N_1$, $N_2N_1N_3$, $N_2N_1 \bullet N_3$, and $N_2 \bullet N_1 \bullet N_3$. The meaning of $N_s \bullet N_t$ in this situation is that N_t is chosen if we observe $z \le p$ for a prescribed $p \in [0,1]$, otherwise N_s is considered. The quantity z is a realisation of Z, a random variable that follows a uniform distribution over $(0,1)$, i.e. $Z \sim U[0,1]$.

We continue by describing how we compute the initial solution in Step 1 of the SRVNS scheme. Therefore, a process plan $PP^0 = \{PP_1^0, \ldots, PP_r^0\}$ is obtained either randomly or the corresponding bills of materials are chosen based on RPT ranking applying the p_{iz} probabilities. Moreover, the routes belonging to these bills of material are chosen based on a discrete uniform distribution. After the bill of materials and the routes are chosen, we can solve the resulting scheduling problem by the LS heuristic described by Sobeyko and Mönch (2016). The notation $LS(w)$ is used to state that the LS heuristic performs w iterations. This heuristic for computing an initial solution is abbreviated by H_{init}.

The shaking step is parallelised to reduce the disadvantage of having only a RVNS approach. Worse solutions in comparison to the incumbent solution are accepted in a manner that is similar to threshold accepting using the setting $\rho(x, x^*) := TWT(x^*) - \frac{1}{\lambda} TWT(x)$. This means that the difference of the TWT value of x^*, the best solution found so far, and the TWT value of x, the incumbent solution, multiplied with a scaling parameter, serves as a threshold. The parallel SRVNS approach tailored in this way is called TPSRVNS. The proposed TPSRVNS scheme can be stated as follows where PP and PP^* are the incumbent process plan and the best process plan found so far, respectively:

TPSRVNS

1. Initialisation: select neighbourhood structures $\tilde{N}_k, k = 1, \ldots, k_{max}$. Compute an initial solution using H_{init}. Initialise $PP^* \leftarrow PP^0$, $PP \leftarrow PP^0$, $TWT^* \leftarrow TWT^0$. Set $k \leftarrow 1$.

2. Repeat the following steps until $iter_{max}$ iterations are reached.

3. Shaking: generate randomly n_{expl} process plans $PP^i \in \tilde{N}_k(PP)$, such that $PP^i \ne PP^j, i \ne j$. For each of the process plans $PP^i, i = 1, \ldots, n_{expl}$, formulate and solve the corresponding flexible job shop scheduling problem. The resulting performance measure values are $TWT^i, i = 1, \ldots, n_{expl}$. Select the process plan PP' with the smallest TWT value.

4. Improvement or not: if $TWT' < TWT^*$ then update $TWT^* \leftarrow TWT'$, $PP^* \leftarrow PP'$.

5. Move or not: if $TWT' < \beta TWT^*$ then update $PP \leftarrow PP'$, $TWT \leftarrow TWT'$, $k \leftarrow 1$. Otherwise, update $k \leftarrow k \mod k_{max} + 1$.

The number of elements in the neighbourhood that are considered in parallel in the shaking step is n_{expl}. The parameter $\beta \geq 1$ is used to control the acceptance of worse solutions. When $\beta = 1$ the TPSRVNS scheme accepts only better solutions. Many new solutions will be accepted if large values of β are chosen. To avoid this undesirable behaviour, we choose β as

$$\beta := 1 + \alpha \exp\left\{1 - iter_{\max}/(iter_{\max} - i)\right\}. \tag{4.4}$$

Here, the current and the maximum number of iterations are i and $iter_{\max}$, respectively. The acceptance rate of worse solutions at the beginning of the algorithm is defined by the parameter α. We obtain $\beta = 1$ for $i = iter_{\max}$. We observe a decreasing rate of acceptance until $iter_{\max}$ iterations are carried out, i.e., it is not expected that worse solutions are accepted at final iterations.

The performance of each newly selected process plan in Step 3 must be evaluated by computing a solution of the corresponding instance of the flexible job shop scheduling problem. Due to the fact that we must solve many of these large-sized instances, heuristics are required that are fast but at the same time also able to provide solutions with small *TWT* values. Based on the computational experiments from Sobeyko & Mönch (2016), we know that the LS procedure causes only modest computing times while at the same time computing solutions of large quality. Although the SBH-type procedures proposed in Sobeyko & Mönch (2016) provide slightly better solutions under some experimental conditions, they cause a large computational burden. The LS approach is therefore chosen in the proposed TPSRVNS approach on the scheduling level for evaluating the quality of the process plans.

It might still lead to a high computational burden when a large number of iterations are performed in TPSRVNS. Since the search space of the TPSRVNS procedure is large in situations including a large number of products, often several hundreds of flexible job shop scheduling problem instances must be performed to determine a process plan of high quality. But at the same time, it is not required to compute schedules of high quality for process plans of low quality. It is expected that low-quality process plans are chosen right after the search process is started.

Hence, we use rules which decide how well a scheduling problem instance must be solved at the current state of the search process. High-quality schedules must be determined after the TPSRVNS procedures reaches promising regions of the search space, otherwise high-quality process plans are ignored. An initial schedule is provided for the LS scheme that is the schedule with the smallest *TWT* value determined by list scheduling using well-known dispatching rules among them the First In First Out (FIFO), Operational Due Date (ODD), Modified Operation Due Date (MOD), and the Weighted Shortest Processing Time (WSPT) rule, and the more advanced global Apparent Tardiness Cost (ATC) dispatching rule with carefully chosen values for the look-ahead parameter κ (cf. Mönch & Zimmermann 2004).

TABLE 4.1

Scheduling Rules Used in the Computational Experiments

Percentage	Scheduling Heuristic
[0–90]	• LS(10.000)
[90–100]	• LS(20.000)
[100–100]	• LS(200.000)

This list scheduling approach is explained in more detail in Sobeyko & Mönch (2016). This multi-pass scheduling algorithm is abbreviated by combined scheduler. If only the simple dispatching rules are used in a multi-pass manner, the resulting approach is called SRD.

Next, we briefly describe the scheduling rules. The notation $[l - u] : H$ is used to express that the scheduling heuristic H is applied after l and before $u\%$ of the $iter_{max}$ iterations in the TPSRVNS procedure. $[l - u] : H$ refers to the situation that H is even used for $u\%$ of the $iter_{max}$ iterations. In the simulation experiments described later in Section 4.6, we will use the rules that can be found in Table 4.1. Recall that the notation LS(k) refers to the situation where k iterations of the LS scheme are applied in Table 4.1. More rules can be found in Sobeyko & Mönch (2016).

4.4.3 Reference Heuristic

The straightforward RND-PP and RPT-PP heuristics are considered that are related to H_{init} (see Section 4.4.2). The heuristics compute n_{ref} process plans. RND-PP randomly chooses process plans, while the process plans are selected in RPT-PP based on RPT ranking as introduced in Section 4.4.2. Scheduling based on the LS approach is performed for these process plans. The process plan with the smallest TWT value is chosen to be executed.

4.5 Simulation-Based Performance Assessment of the IPPS Approach

4.5.1 Simulation Infrastructure

The infrastructure for simulation-based performance assessment replaces the shop floor in Figure 4.1 by a simulation model of the manufacturing system following the discrete-event paradigm. It extends the simulation infrastructure for production planning and control proposed by Mönch et al. (2003), Mönch (2007), and Ponsignon and Mönch (2014) towards process planning.

The simulation model mimics the behaviour of the machines and orders, the so-called base system of the manufacturing system (cf. Mönch et al. 2013). The backbone of the infrastructure is a data layer, also known as blackboard, in the memory of the simulation computer. It contains the following major business objects from the base system and process:

- the machines with their current status
- the orders with their current status
- the bills of materials selected for the different products
- the routes that belong to a certain bill of materials
- information related to the schedules that have to be executed on the shop floor.

The overall architecture is shown in Figure 4.2.

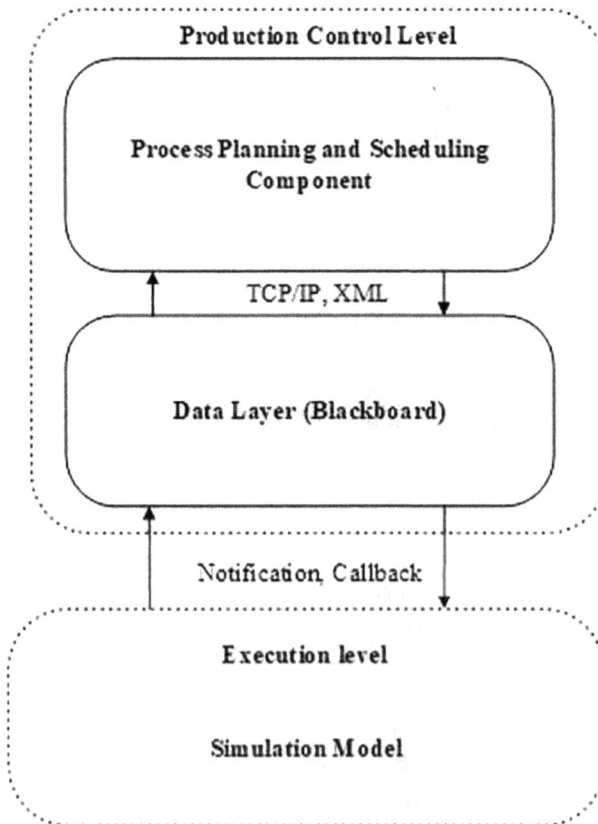

FIGURE 4.2
Simulation infrastructure.

The process planning and scheduling component and the data layer are able to exchange information in both directions. Note that the process planning and scheduling component in principle can be divided into two components.

Unlike the communication between the simulator and the data layer, the communication between the process planning and scheduling component and the data layer is implemented based on the transmission control protocol/internet protocol (TCP/IP) (cf. Fall & Stevens 2011). XML (Bray et al. 1997) messages are used for the communication between the two components. A client-server architecture is implemented for the communication process. The process planning and scheduling component are designed as a server application. It serves as a listener for newly incoming messages. As soon as new data arrives, it is parsed, and the resulting IPPS problem instance is solved. The computed schedule is converted into the XML format and sent back to the data layer.

The data layer acts as a client that exchanges messages with the process planning and scheduling component. As soon as it is decided to trigger process planning and scheduling activities, the relevant data is collected and converted into an XML message. Next, the client tries to establish a connection with the server and sends the message. The communication among client and server is synchronous, i.e., the connection is kept open and the client waits until the server sends back a new schedule. As soon as this happens, the connection closes, the received XML message is parsed, and the schedule becomes available on the shop floor.

The data layer is implemented as an extension of the commercial simulation engine AutoSched AP in form of a Dynamic Link Library (DLL). It communicates with the simulation software by means of notification functions. The client used for communication with the server is implemented in form of a DLL as well. It provides interface functions which can be called by the data layer to trigger the IPPS process. The client is based on the TCP/IP technology available in the Qt 4.8.5 software library.

All algorithms described in this chapter as well as the complete infrastructure functionality are developed by applying the C++ programming language. The simulation engine AutoSched AP 9.3 is also a C++ class library.

4.5.2 Rolling Horizon Scheme

A rolling horizon approach is taken to deal with a dynamic and uncertain manufacturing environment. Therefore, we introduce the concept of process planning and scheduling epochs. At each epoch, an IPPS problem instance is formed based on the data from the base system. We denote the time span between two consecutive epochs by t_Δ. This means that IPPS instances are solved at the time points $t_k = kt_\Delta, k = 0,1,\ldots$. A process planning and scheduling window with length

$$t_h = t_\Delta + t_{ah} \qquad (4.5)$$

is considered. Here, the quantity t_{ah} is called the additional horizon. Note that only those orders are considered when the IPPS instance in epoch k is formed that have a ready time that is smaller than $t_k + t_h$. The taken rolling horizon approach can be summarised as follows:

IPPS Rolling Horizon Approach

1. Initialise $k \leftarrow 0, t_k \leftarrow 0, t_h = t_\Delta + t_{ah}$.
2. Repeat until the end of the simulation horizon is reached.
3. Consider the current state of the manufacturing system, especially of the machines and the work in progress (WIP) orders.
4. Consider all non-WIP orders with a ready time smaller than $t_k + t_h$.
5. Solve the IPPS problem instance formed in Steps 3 and 4.
6. Update $t_{k+1} \leftarrow t_k + t_\Delta$.
7. Implement the obtained schedule for the period $[t_k, t_{k+1}]$.
8. Update $k \leftarrow k + 1$.

Note that each IPPS problem instance is solved in Step 5 for all the remaining operations. However, since only the schedule is implemented in a period of length t_Δ it is possible to consider scheduling problem instances where the number of scheduled operations is determined by the length of the process planning and scheduling window. Of course, this approach requires setting appropriate internal due dates (see Mönch et al. 2007, Mönch & Zimmermann 2011) which is a non-trivial task. But since the size of this scheduling problem instance is much smaller for a scheduling window of length t_h it can be solved using a much smaller amount of computing time.

4.6 Simulation Experiments

4.6.1 Design of Experiments

We consider a manufacturing scenario with five different products. Fifteen machines exist in the considered manufacturing system. These machines are divided into five machine groups where two groups contain four machines each, two groups have three machines per group, and one machine group has only a single machine.

The length of the simulation horizon is 15 days. Around 750 orders arrive during this period of time. Orders that are already known before the simulation is started have ready times that follow a discrete uniform distribution over the set of integers $\{0, 10, 20, 30\}$. The inter arrival time of the remaining orders is selected as 180 min. This setting leads to a flow factor (FF) between 1.5 and 2.0. The FF is defined as the ratio of the difference of the completion and the ready time of an order and the RPT. The due dates are set according to the FF-based approach described in Sobeyko & Mönch (2016). Tight due dates are used in the simulation experiments. Moreover, we set $t_\Delta = 120$ min and $t_{ah} \in \{0, 60\}$ min.

The IPPS approach selects the bill of materials and routes for each product. In a rolling horizon approach, this leads to the following settings:

1. Preservation of the already chosen bill of materials and routes for each order: if a bill of materials and the corresponding routes are selected for a specific order in a process planning and scheduling epoch, then this setting cannot be changed anymore in future planning epochs. This leads to a situation where different bills of materials and routes might exist for the same product at the same time.

2. Flexible choice of the bills of materials and routes for each order: two cases have to be differentiated.

   a) If no subparts of an order have been started with processing, i.e., if the order is not executed in the simulation model then the order can undergo process planning and scheduling in the new process planning and scheduling epoch, as depicted in Figure 4.3. Note that this is always the situation for new orders which are not planned yet.

   b) If at least one subpart of a certain order has been started with processing, i.e., if the order is already executed, then the remaining subparts can obtain new alternative routes as depicted in Figure 4.4. These orders can be rescheduled in the new process planning and scheduling epoch. Note that this again might lead to a situation where different bills of materials and routes might exist for the same product at the same time.

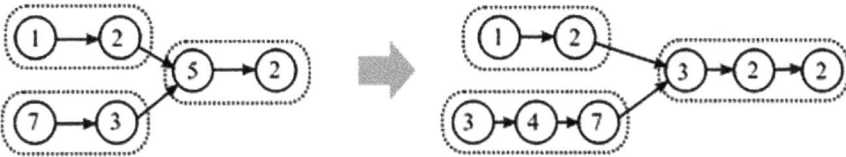

FIGURE 4.3
Changing the entire process plan is allowed.

FIGURE 4.4
Only changing the routing of subparts is allowed.

The former situation is abbreviated by Preservation (P) and the latter one by No Preservation (NP).

We perform five independent solution replications for each process planning and scheduling instance since the VNS approach and the LS procedure have stochastic ingredients. The schedule with the lowest *TWT* value is selected to be executed in the simulation model. Instead of using the operation start times from the schedule, we sort the operations in non-decreasing order with respect to their start times and always choose the operation with the highest priority when a machine becomes available.

The experimental design is summarised in Table 4.2. We consider only a fractional design of experiments, namely five different scenarios. They are shown in Table 4.3.

The average *TWT* value of all orders completed within the simulation horizon is reported. Moreover, we are interested in the average cycle time (*CT*) of all completed orders. The *CT* of an order O_{is} is defined as $CT_{is} := C_{is} - r_{is}$. The average *CT* is taken over all products and orders.

The simulation experiments are conducted on a computer with an Intel Core i7-3630QM processor and 20 GB of main memory.

TABLE 4.2

Design of Experiments

Factors	Level	Count
Process planning approach (PPA)	RPT-PP, TPSRVNS	2
Scheduling approach (SA)	SRD, LS(200.000), scheduling rules from Table 4.1	3
Additional scheduling horizon (ASH)	$t_{ah} \in \{0, 60\}$	2
Preservation of the process plans	P, NP	2
Number of independent replications of the VNS scheme per instance	5	

TABLE 4.3

Features of the Different Scenarios

Scenario	PPA	SA	ASH	Preservation of the Process Plans
S1	RPT-PP	SRD	$t_{ah} = 0$	Not applicable
S2	RPT-PP	LS(200.000)	$t_{ah} = 0$	Not applicable
S3	TPSRVNS	Scheduling rules from Table 4.1	$t_{ah} = 0$	P
S4	TPSRVNS	Scheduling rules from Table 4.1	$t_{ah} = 0$	NP
S5	TPSRVNS	Scheduling rules from Table 4.1	$t_{ah} = 60$	P

4.6.2 Parameter Setting

In the simulation experiments, we consider the neighbourhood structure $N_2N_1N_3$ on the process planning level. The number of different products is chosen as value of k_{max}. The abbreviation $\tilde{N}_k := N_2N_1N_3(k)$ is applied. We select $iter_{max} = 100$ and $n_{expl} = 4$ for the TPSRVNS algorithm. The acceptance rate $\alpha = 0.15$ is selected in (4.4).

These settings are justified by the computational experiments in Sobeyko & Mönch (2016). The flexible job shop scheduling approach is parameterised according to the settings found in Table 4.1, i.e., the scheduling rules are used within the different iterations.

4.6.3 Simulation Results

We show the simulation results, namely the average *TWT* values and the average *CT* values, in Table 4.4. The performance measure values are reported relative to those for scenario S2.

We see from Table 4.4 that the RPT-PP approach for process planning in combination with the SRD list scheduling approach is outperformed by the RPT-PP approach combined with the LS(200.000) scheduling approach. Therefore, scenario S2 is used as reference. Large *TWT* reductions are also possible by the TPSRVNS approach described in Section 4.4.2. Scenario S4 leads to a *TWT* reduction of up to 52%. This behaviour is caused by the flexibility to change the bills of materials and routes when a new process planning and scheduling epoch starts. The *TWT* value of the S5 scenario is worst compared to S3 and S4 due to the longer horizon for process planning and scheduling, in combination with the inability to change the bills of materials and routes for orders that are already in execution. Note that some minor *CT* improvements are possible when the TPSRVNS approach is applied.

Overall, we clearly see that the advantage of TPSRVNS-type approaches over the reference heuristic observed for static problem instances in Sobeyko & Mönch (2016) carries over to the dynamic setting. However, it is important to parameterise the rolling horizon scheme in an appropriate manner.

TABLE 4.4

Simulation Results

Scenario	TWT	CT
S1	1.31	1.01
S2	1.00	1.00
S3	0.55	0.97
S4	0.48	0.97
S5	0.70	1.01

4.7 Conclusion and Future Research Directions

An approach to solve an IPPS problem was described in this chapter. The corresponding shop floor was given by a large-scale flexible job shop. The *TWT* performance measure was used. We proposed a two-layer hierarchical approach to tackle this problem. A VNS-based scheme was used on the top level to choose bills of materials and routes per product. On the base level, the resulting large-sized flexible job shop scheduling problem instances were solved by an LS approach. A disjunctive graph representation of the flexible job shop was used. We described a simulation infrastructure to assess the performance of the IPPS approach at hand in a rolling horizon setting using discrete-event simulation. The implementation of the IPPS prototype is briefly sketched. Results of the simulation experiments were discussed that clearly demonstrate that certain configurations of the IPPS approach are more beneficial than others when the IPPS approach is executed in a rolling horizon setting. The advantage of TPSRVNS-type approaches compared to simple reference heuristics observed for static problem instances in Sobeyko & Mönch (2016) carries over to the dynamic setting when the rolling horizon scheme is parameterised in an appropriate way.

The following future research directions are possible. It seems to be desirable to extend the simulation study presented in the present chapter by including machine failures and stochastic processing times and release dates of the orders. Moreover, simulation studies with larger flexible job shops are very interesting.

Another avenue of future research is given by studying possibilities to compute robust process plans and schedules in the sense that they are already buffered against several uncertainties from the shop floor and the order arrival process. This requires that simulation is used to evaluate the quality of process plans and schedules in the presence of machine breakdowns and processing time and order arrival uncertainty.

Moreover, we believe that it is desirable to use the proposed IPPS scheme in holonic multi-agent systems (MAS) that follow the Product Resource Order Staff Architecture (PROSA) (Van Brussel et al. 1998, Van Belle et al. 2012). The process planning functionality, namely the VNS approach, could be integrated into staff agents that belong to the product agents while the necessary scheduling functionality could be part of a staff agent for scheduling. The implementation of such a MAS prototype can be based on the ManufAg framework proposed by Mönch & Stehli (2006).

References

Adams, J., Balas, E., Zawack, D. 1988. The shifting bottleneck procedure for job shop scheduling. *Management Science*, 34(3), 391–401.

Barua, A., Narasimhan, R., Upasani, A., Uzsoy, R. 2005. Implementing global factory schedules in the face of stochastic disruptions. *International Journal of Production Research*, 43(4), 94–109.

Brandimarte, P., Calderini, M. 1995. A hierarchical bicriterion approach to integrated process plan selection and job shop scheduling. *International Journal of Production Research*, 33(1), 161–181.

Bray, T., Paoli, J., Sperberg-McQueen, C.-M., Maler, E., Yergeau, F. 1997. Extensible markup language (XML). *World Wide Web Journal*, 2(4), 27–66.

Driessel, R., Mönch, L. 2012. An integrated scheduling and material handling approach for complex job shops: a computational study. *International Journal of Production Research*, 50(20), 5966–5985.

Fall, K. R., Stevens, W. R. 2011. *TCP/IP Illustrated, Vol. 1: The Protocols*, 2nd edition. Addison-Wesley, Boston, MA.

Hansen, P., Mladenovic, N. 2001. Variable neighborhood search: principles and applications. *European Journal of Operational Research*, 130, 449–467.

Hansen, P., Mladenovic, N. 2003. Variable neighborhood search. In *Handbook of Metaheuristics*, F. Glover and G. A. Kochenberger (eds.), pp. 145–184. Springer, Boston, MA

Ivens, P., Lambrecht, M. 1996. Extending the shifting bottleneck procedure to real-life applications. *European Journal of Operational Research*, 90, 252–268.

Lawler, E. L. 1977. A "pseudopolynomial" time algorithm for sequencing jobs to minimize total tardiness. *Annals of Discrete Mathematics*, 1, 331–342.

Lee, H., Kim, S.-S. 2001. Integration of process planning and scheduling using simulation based genetic algorithms. *International Journal of Advanced Manufacturing Technology*, 18(8), 586–590.

Mönch, L. 2007. Simulation-based benchmarking of production control schemes for complex manufacturing systems. *Control Engineering Practice*, 15, 1381–1393.

Mönch, L., Drießel, R. 2005. A distributed shifting bottleneck heuristic for complex job shops. *Computers & Industrial Engineering*, 49(3), 363–380.

Mönch, L., Fowler, J. W., Mason, S. J. 2013. *Production Planning and Control for Semiconductor Wafer Fabrication Facilities: Modeling, Analysis and Systems*, Vol. 52. Springer Operations Research/Computer Science Interfaces, New York.

Mönch, L., Rose, O., Sturm, R. 2003. A simulation framework for the performance assessment of shop-floor control systems. *Simulation*, 79(3), 163–170.

Mönch, L., Schabacker, R., Pabst, D., Fowler, J. W. 2007. Genetic algorithm-based subproblem solution procedures for a modified shifting bottleneck heuristic for complex job shops. *European Journal of Operational Research*, 177(3), 2100–2118.

Mönch, L., Stehli, M. 2006. ManufAg: a multi-agent-system framework for production control of complex manufacturing systems. *Information Systems and e-Business Management*, 4(2), 159–185.

Mönch, L., Zimmermann, J. 2004. Improving the performance of dispatching rules in semiconductor manufacturing by iterative simulation. Proceedings of the 2004 Winter Simulation Conference, 2, 1881–1887.

Mönch, L., Zimmermann, J. 2011. A computational study of a shifting bottleneck heuristic for multi-product complex job shops. *Production Planning & Control*, 22(1), 25–40.

Ovacik, I. M., Uzsoy, R. 1997. *Decomposition Methods for Complex Factory Scheduling Problems*. Kluwer Academic Publishers, Boston, MA.

Pfund, M. E., Balasubramanian, H., Fowler, J.W., Mason, S.J., Rose, O. 2008. A multi-criteria approach for scheduling semiconductor wafer fabrication facilities. *Journal of Scheduling*, 11, 29–47.

Phanden, R. K., Jain, A., Verma, R. 2011. Integration of process planning and scheduling: a state-of-the-art review. *International Journal of Computer Integrated Manufacturing*, 24(6), 517–534.

Phanden, R. K., Jain, A., Verma, R. 2013. An approach for integration of process planning and scheduling. *International Journal of Computer Integrated Manufacturing*, 26(4), 284–302.

Ponsignon, T., Mönch, L. 2014. Simulation-based performance assessment of master planning approaches in semiconductor manufacturing. *OMEGA*, 4(6), 21–35.

Sobeyko, O., Mönch, L. 2016. Heuristic approaches for scheduling jobs in large-scale flexible job shops. *Computers & Operations Research*, 68, 97–109.

Sobeyko, O., Mönch, L. 2017. Integrated process planning and scheduling for large-scale flexible job shops using metaheuristics. *International Journal of Production Research*, 55(2), 392–409.

Upasani, A. A., Uzsoy, R. 2008. Integrating a decomposition procedure with problem reduction for factory scheduling with disruptions: a simulation study. *International Journal of Production Research*, 46(21), 5883–5905.

Van Belle, J., Philips, J., Ali, O., Germain, B. S., Van Brussel, H., Valckenaers, P. 2012. A service-oriented approach for holonic manufacturing control and beyond. *Service Orientation in Holonic and Multi-Agent Manufacturing Control, Studies in Computational Intelligence*, 402(1), 1–20.

Van Brussel, H., Wyns, Valckenaers, P., Bongaerts, L., Peeters, P. 1998. Reference architecture for holonic manufacturing systems: PROSA. *Computers in Industry*, 37(3), 255–274.

Wong, T. N., Leung, C. W., Mak, K. L., Fung, R. Y. K. 2006. An agent-based negotiation approach to integrate process planning and scheduling. *International Journal of Production Research*, 44(7), 1331–1351.

Zhang, R., Ong, S. K., Nee, A. Y. C. 2015. A simulation-based genetic algorithm approach for remanufacturing process planning and scheduling. *Applied Soft Computing*, 37, 521–532.

Ziarnetzky, T., Kacar, B., Mönch, L., Uzsoy, R. 2015. Simulation-based performance assessment of production planning formulations for semiconductor wafer fabrication. *Proceedings of the 2015 Winter Simulation Conference*, pp. 2884–2895.

5

Integration of Process Planning and Scheduling in an Energy-Efficient Flexible Job Shop: A Hybrid Moth Flame Evolutionary Algorithm

Vijaya Kumar Manupati and Subhash C. Bose
National Institute of Technology Warangal

Rajeev Agrawal
Malaviya National Institute of Technology Jaipur

Goran D. Putnik and M.L.R. Varela
University of Minho

CONTENTS

5.1 Introduction 116
5.2 Literature Review 117
    5.2.1 Job Shop Scheduling Problem (JSSP) 117
    5.2.2 FJSSP 117
5.3 Problem Description 118
    5.3.1 Objectives 119
    5.3.2 Subject to Constraints 119
5.4 The NSGA-II and Proposed Multi-Objective Hybrid Moth
    Flame Evolutionary Algorithm (MOO-HMFA) 120
    5.4.1 NSGA-II 120
        5.4.1.1 Initial Population Generation 120
        5.4.1.2 Operators in NSGA-II 121
    5.4.2 Hybrid Moth Flame Optimisation (MFO) Algorithm 121
5.5 Experimental Evaluation 125
5.6 Discussion on Experimental Results 127
5.7 Conclusions 131
References 132

5.1 Introduction

Rapid developments in the manufacturing sector gain the advantage by shifting its phase from a traditional manufacturing approach to advanced manufacturing concepts. Subsequently, recent stringent government policies for protecting the environment from emissions are one of the objectives of the current approach (Tang et al., 2016). However, by implementing the recent approaches in order to achieve cost-effective, environmentally friendly, and efficient manufacturing processes, the current manufacturing systems need to be flexible enough and highly efficient. However, due to the increase in the price of unit energy and stringent environmental protection awareness schemes, meeting the mentioned requirements becomes a challenge. However, out of the mentioned objectives, reducing the energy consumption of manufacturing process has become a critical problem, thus recently many researchers drew their attention towards it (Jiang et al., 2014). FJSSP has several applications in various industries such as factories of plastic injection machines, fastener manufacturing companies, etc. To investigate the above-mentioned sustainable parameters in FJSSP, the objective functions such as energy consumption, completion time, and processing cost have been considered for sustainable manufacturing.

FJSSP usually deals with real-time manufacturing situations where nature is non-deterministic polynomial-time hard (NP-hard) and more complex than job shop scheduling problem (JSSP) (Jiang et al., 2014). As the arrival of jobs is dynamically varying and high in number, finding the optimal sequence in real time is a complicated task. To arrive at the optimal sequence, in general, flexible job shop scheduling problem (FJSSP) can be solved using two approaches viz., hierarchical and integrated (Brandimarte, 1993). But with this approach, it is difficult to attain the best possible solutions and requires a compromise with near-optimal solutions. The considered problem is difficult to address using the hierarchical approach due to the demand of multiple optimal solutions. Thus, efficient optimisation algorithms with outstanding computational effort are required to deal with multiple objective functions. We focus on responding to the following questions.

1. What kind of mathematical model must be developed with the considered performance measures, and how this model can be solved to optimality?
2. What are the effects of the proposed multi-objective evolutionary algorithm, i.e., multi-objective optimisation hybrid moth flame (MOO-HMF) with IPPS approach?
3. How do these three conflicting objectives trade-off?

5.2 Literature Review

The discussion here will focus on the three identified production scheduling problems and their detailed literature is mentioned as follows.

5.2.1 Job Shop Scheduling Problem (JSSP)

Fang et al. (1993) proposed a new and promising genetic algorithm (GA) approach for finding the makespan of JSSP. Pezzella and Merelli (2000) developed a new variant of Tabu search technique for JSSP which is guided by shifting the bottleneck model that considered makespan as the objective to attain the best solution. Gonçalves et al. (2005) addressed the job-shop production environment problem and solved it using the proposed hybrid GA approach for minimising the maximum completion of time. Whereas Liu et al. (2014) solved JSSP by minimising the non-processing energy consumption and weighted tardiness. In, May et al. (2015) a multi-objective genetic algorithm (MOGA) approaches considering makespan and energy consumption of a JSSP has been developed using a green GA. Zhang et al. (2016) developed the energy requirement and weighted tardiness of the JSSP and derived the exact solution by using the GA. Yin et al. (2017) formulated a multi-objective optimisation model that considers "productivity, energy efficiency and noise reduction", for a JSSP and proposed a simple lattice design-based MOGA for optimal solutions.

5.2.2 FJSSP

Due to the real-life applications of FJSSP, simultaneous optimisation of multiple objectives needs to be done. Therefore, many researchers have started working on multi-objective optimisation of FJSSP (Reddy et al., 2017; Silva et al., 2017; Varela et al., 2017; and Santos et al., 2015). Gen et al. (2007) built a multi-objective optimization problem (MOOP) considering makespan, maximum machine workload, and total workload as objectives for a FJSSP. A hybridised GA has been developed to solve it for acquiring accurate results. Zhang et al. (2009) addressed a MOP-related to "completion time, the workload of the critical machine and total workload of machines", in a FJSSP to obtain the most optimal results based on hybrid particle swarm optimization (PSO). Gen et al. (2009) used a GA with crossover and mutation operations for FJSSP with the objectives of "minimising makespan, total workloads of machines and maximum workloads of machines", to obtain accurate solutions. Madureira et al. (2010) developed a BioSched system to support the dynamic and distributed scheduling of extended JSSP. Li et al. (2011) developed a Pareto-based discrete ant bee colony (ABC) algorithm to elucidate MOFJSSP to obtain best results. Jiang et al. (2014) developed a MOFJSSP involving "makespan, processing cost, energy consumption, and cost-weighted processing quality", with the proposed non-dominated sorting genetic algorithm (NSGA) for attaining the best results. Madureira et al. (2014) proposed a negotiation mechanism for the extended JSSP that considers makespan and machine occupation rate as objective functions. Karthikeyan et al. (2015) proposed a hybrid discrete firefly algorithm for the FJSSP that considers "maximum completion time, the workload of a critical machine and total workload of all machines", to obtain the results.

Although a plethora of research studies have been conducted on multi-objective based FJSSP, to the author's knowledge, till date very few works of literature are available with the consideration of multiple objectives such as energy consumption of machines, processing cost, and makespan in the context of FJSSP. In this chapter, by considering the above-mentioned objective functions in the context of FJSSP, a multi-objective mathematical model along with constraints has been developed. To obtain good approximate solutions, we proposed a MOO-HMF algorithm. The experiments are conducted using complex scenarios with job complexities to assess the performance of the proposed hybrid multi-objective algorithm. Later, a comparison study of the proposed algorithm's effectiveness and efficiency with NSGA-II, in order to obtain good approximate solutions, is conducted. In later sections, the problem description, proposed framework, and its experimentation have been detailed.

Section 5.3 details the description of the considered problem and its representation with a mathematical model along with the constraints. In Section 5.4, the framework of the proposed MOO-HMF algorithm has been described. In Section 5.5, analysis of different instances has been discussed in detail in Section 5.6. Finally, the paper concludes with Section 5.7.

5.3 Problem Description

We addressed a FJSSP with a "set of n jobs and v_i as a number of operations varying between $(O_{j1}, O_{j2}, ..., O_{jl})$. Each operation O_{jl} of job j has to be processed on machine k from a set of eligible machines m". It is assumed that the operations' processing time is known in advance and the processing machines are available at time zero. The machines must process only one operation at a time while the subsequent jobs wait at the buffer stations which are infinite space until the operation finishes. In this chapter, to understand the system performance. makespan is considered for evaluation. Thereafter, the sustainability parameters such as power and time have been considered due to the fact that the speed of a machine is directly proportional to its energy consumption despite its shorter processing time. Therefore, the performance measure, i.e., minimisation of energy consumption, has been considered as a second objective function. In addition to the above performance measures, minimisation of the total processing cost is also considered. Due to the nature of the problem, a mathematical model is shown in Table 5.1 below with notations.

TABLE 5.1

Notations Used in the Mathematical Model

n	The total number of jobs
M	Machines available
v_j	Number of operations of job j
C_{jo}^k	Completion time of oth operation of job j on machine k
T_{jo}^k	Duration of oth operation of job j on machine k
S_{jo}^k	Beginning time of oth operation of job j on machine k
mc_j	Denotes raw material cost of job j
pc_k	Denotes the process cost of machine k per unit time
e_k	Denotes the energy consumption of machine k per unit time

5.3.1 Objectives

In this problem, minimisation of maximum total completion time, total processing cost minimisation, and minimisation of energy consumption are the three objectives and their formulation is shown in the following equations:

$$\text{Minimization of Total completion time } (T) = \text{Max}\left(T_{jo}^k\right) \tag{5.1}$$

$$\text{Total processing cost minimization } (TPC) = \sum_{j=1}^{n} mc_j + \sum_{j=1}^{n}\sum_{o=1}^{v_i} O_{jo}^k T_{jo}^k pc_k \tag{5.2}$$

$$\text{Minimization of energy consumption } (E) = \sum_{j=1}^{n}\sum_{o=1}^{v_i} O_{jo}^k T_{jo}^k e_k \tag{5.3}$$

5.3.2 Subject to Constraints

$$T_{jo}^k \geq T_{j(o-1)}^k \tag{5.4}$$

$$T_{jo}^k \geq T_{(j-1)o}^k \tag{5.5}$$

$$T_{jo}^k \geq 0 \tag{5.6}$$

Equation (5.1) represents the primary objective of the problem, i.e., minimisation of makespan; to minimise the total processing cost of operations which is mentioned in equation (5.2), and to minimise the energy consumption of machines which is mentioned in equation (5.3). The constraints are in the equations (5.4)–(5.6). Equations (5.4) and (5.5) show how the precedence constraints between the operations and jobs are not violated, i.e., the oth operation only starts after $(o-1)$th operation finishes its process, similar to the case of the jth job. Equation (5.6) indicates the preceding constraints of operations.

5.4 The NSGA-II and Proposed Multi-Objective Hybrid Moth Flame Evolutionary Algorithm (MOO-HMFA)

Hence, the considered problem is NP-hard in nature, where it cannot be solved in a polynomial amount of time, and there is a need for optimisation techniques/methods/algorithms. Among the many algorithms available in the market, the nature-inspired evolutionary algorithms have an advantage due to their merit. In addition, to improve the accuracy of the solution a MOO-HMFA has been proposed and it has been discussed in detail in this section.

5.4.1 NSGA-II

In this chapter, first NSGA-II was adopted from Deb et al. (2002a) and applied to the considered problem for generating the optimal process plans.

5.4.1.1 Initial Population Generation

The algorithm starts with the initial population generation having size as random in order. Later, a job is nominated at random and verified against its predecessor constraint. After the predecessor constraint is satisfied, the chromosome is defined by adding the variables in the bits. Otherwise, a new job is randomly nominated and verified for its predecessor constraint. Until all the chromosomes are completely occupied with the jobs, the process continues. For example, in Figure 5.1 the encoding scheme of the chromosome is clearly depicted, and the array shows the operations performed by the machines. Here, the array size is maintained as 14 with different operations randomly allocated. Similarly, machines are arranged in second array processing cost and the energy consumption of machines as corresponding arrays, respectively.

1	2	3	4	5	6	7	8	9	10	11	12	13	14
O_{41}	O_{21}	O_{42}	O_{23}	O_{43}	O_{31}	O_{33}	O_{22}	O_{11}	O_{13}	O_{44}	O_{12}	O_{32}	O_{14}
m_2	m_3	m_3	m_3	m_2	m_2	m_2	m_2	m_4	m_3	m_2	m_4	m_3	m_3
pc_2	pc_3	pc_3	pc_3	pc_2	pc_2	pc_2	pc_2	pc_4	pc_3	pc_2	pc_4	pc_3	pc_3
e_2	e_3	e_3	e_3	e_2	e_2	e_2	e_2	e_4	e_3	e_2	e_4	e_3	e_3

FIGURE 5.1
Chromosome representation for NSGA-II.

5.4.1.2 Operators in NSGA-II

Operators of NSGA-II, namely, crossover and mutation for exploration and exploitation, are used for further process, where crossover helps to search in the not-examined solution space so as to generate a solution which will be suggestively different from the past solutions. An appropriate value for the probability of crossover plays a vital role so as to make sure that the performance of the proposed algorithm is maintained. A new population with size 2N is defined by merging the parent and child populations. The combined population is to make sure that there is elitism which produces a huge diversity of the population so as to attain better convergence.

5.4.2 Hybrid Moth Flame Optimisation (MFO) Algorithm

We adopted a new bio-inspired evolutionary algorithm, i.e., MFO algorithm, to find the best possible solutions for the considered problem. The adopted algorithm has several applications in various diverse fields; here, we proposed a new MFO that suits the need of the considered problem, and which covers all possible solution sets to obtain an optimised result.

In Figure 5.2 the flowchart of the proposed MOO-HMFO is depicted where the moths in the algorithm are defined as possible solutions and their positions in space are considered as the variables of the problem. The basic assumption for performing the algorithm is that the moths can fly in all dimensions together and individually, or even in hyper-dimensional space by changing their position vectors.

Step 1. Initially, we define the space for moths to explore in the form of time matrices and their related inputs. An encoding scheme for the initialisation of processing time in the form of a chromosome is shown in Figure 5.3.

Step 2. Next, we multiply the time matrices with our processing cost inputs and energy consumption inputs to get our remaining search spaces as energy and processing cost.

$$P\_cost(j,k,i) = Processing\ cost\ (k) * time\ (j,k,i) \tag{5.7}$$

$$Energy\ (j,k,i) = Energy\_consumption\ (k) * time\ (j,k,i) \tag{5.8}$$

Equations (5.7) and (5.8) represent the processing cost of jobs and their energy consumption on machines. The procedure for the calculation of processing cost and energy consumption matrix is shown in Algorithm 2.

Step 3. We then convert time and energy matrices in terms of money by multiplying the matrices by wages and cost per unit energy and add all three matrices along with their weights. This mutation is applied so as to dimensionally match the matrices of time, cost, and energy.

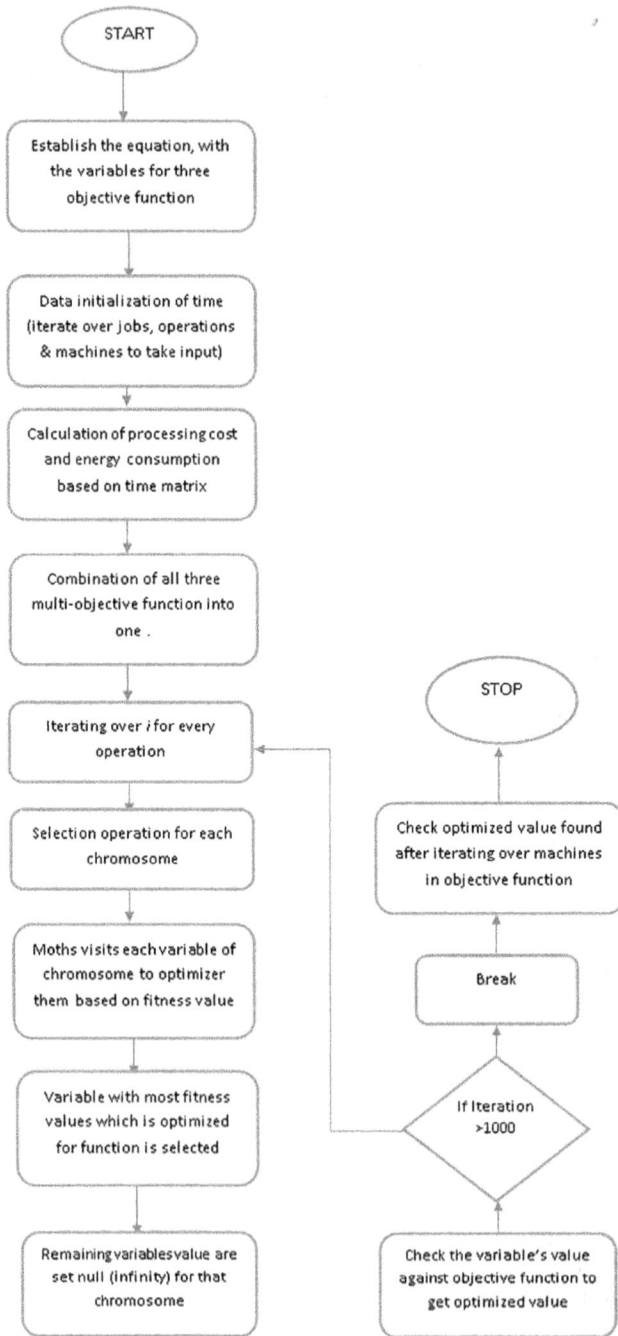

FIGURE 5.2
MOO-HMF detailed flowchart.

Chromosome encoded based on procedure											
4	12	9	14	6	11	5	23	8	13	3	10
	Machine1			Machine2			Machine3			Machine4	

Chromosome encoded based on procedure											
5	7	15	8	4	9	12	3	17	13	9	11
	Operation1			Operation2			Operation3			Operation4	

Chromosome encoded based on procedure											
14	6	11	9	3	12	2	13	19	9	12	4
	Job1			Job2			Job3			Job4	

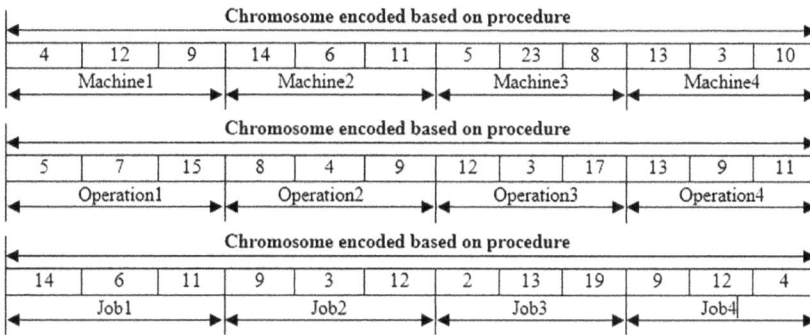

FIGURE 5.3
Encoding scheme of chromosome for initialisation.

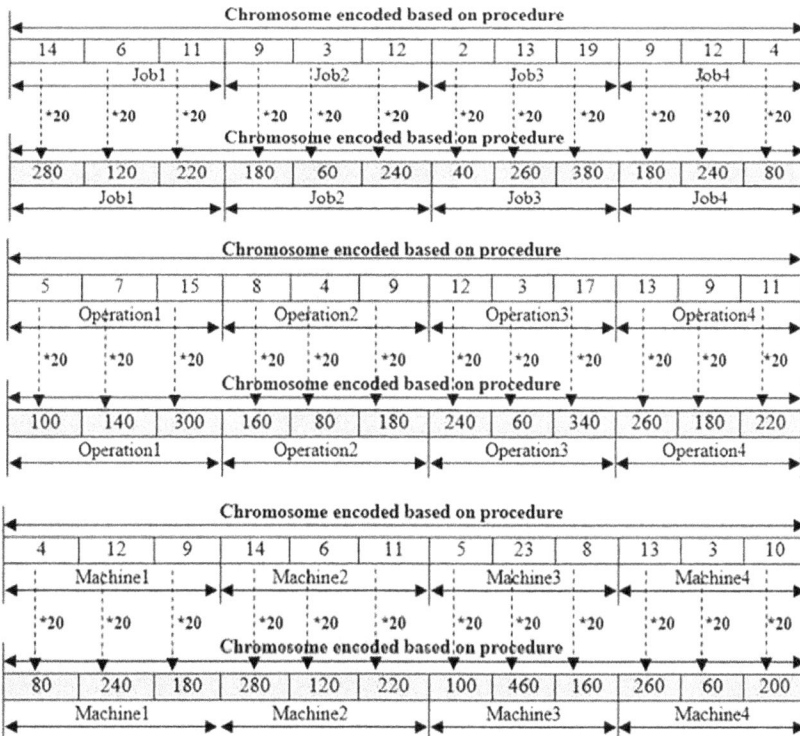

Chromosome encoded based on procedure											
14	6	11	9	3	12	2	13	19	9	12	4
	Job1			Job2			Job3			Job4	
*20	*20	*20	*20	*20	*20	*20	*20	*20	*20	*20	*20

Chromosome encoded based on procedure

280	120	220	180	60	240	40	260	380	180	240	80
	Job1			Job2			Job3			Job4	

Chromosome encoded based on procedure											
5	7	15	8	4	9	12	3	17	13	9	11
	Operation1			Operation2			Operation3			Operation4	
*20	*20	*20	*20	*20	*20	*20	*20	*20	*20	*20	*20

Chromosome encoded based on procedure

100	140	300	160	80	180	240	60	340	260	180	220
	Operation1			Operation2			Operation3			Operation4	

Chromosome encoded based on procedure											
4	12	9	14	6	11	5	23	8	13	3	10
	Machine1			Machine2			Machine3			Machine4	
*20	*20	*20	*20	*20	*20	*20	*20	*20	*20	*20	*20

Chromosome encoded based on procedure

80	240	180	280	120	220	100	460	160	260	60	200
	Machine1			Machine2			Machine3			Machine4	

FIGURE 5.4
Proposed mutation by considering operator wages.

Because of the difference in size of the array and their dimensions a different type of mutation is proposed to form a matrix with minimum values, as shown in Figures 5.4 and 5.5. Later, the matrix is multiplied with the energy by its unit cost and time with operator wages.

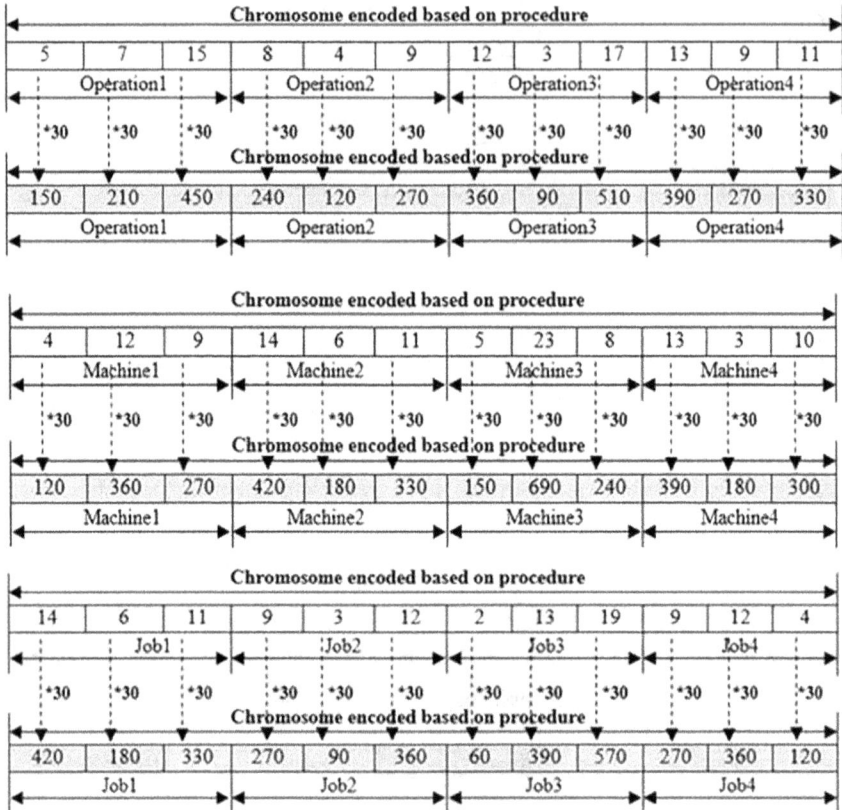

FIGURE 5.5
Proposed mutation by considering the unit cost of energy.

Step 4. All the rows are considered as individual light sources in which the minimum value is to be found.

Step 5. We then start exploring the matrices (search spaces) to find the minimum of the sum so as to obtain a Gantt chart to analyse the results.

Step 6. After the inputs are received and our search area has been properly defined, we then start exploring the matrices row by row, searching for the minimum entry in their respective rows.

Step 7. For the same, we finally find the sum of all the resultant matrices columnwise to gain a single-valued function after converting all ∞s to zeros.

Step 8. We then check whether the single-valued function is optimised or not based on the fitness values to select the highest probable solution for our problem. In this example, as shown in Figure 5.3, we can see that the second chromosome has the best fitness values. This process is repeated to find the optimal solution.

5.5 Experimental Evaluation

The experiments have been performed considering different modified FJSSP benchmark instances, incorporating processing costs and energy consumption profiles for a different set of jobs and machines. First, a FJSSP is proposed and according to our requirements, we constructed ten FJSSP benchmarks. These constructed benchmarks are based on the instances proposed by Brandimarte (1993) and Fisher and Thompson (1963). This is shown in Table 5.2. Thereafter, the processing cost matrix is generated by multiplying the unit processing cost matrix with processing time matrix. Similarly, the energy consumption matrix is formed by multiplying the unit energy consumption matrix with the processing time matrix. It can be inferred from these matrices that, with a decrease in processing time, processing cost decreases and the energy consumption also decreases. All instances considered here are to determine that the results were benchmark instances with different instances $10 \times 6 \times 6$, $15 \times 10 \times 10$, $20 \times 5 \times 5$, $10 \times 10 \times 10$, $10 \times 5 \times 5$, $10 \times 9 \times 9$, $6 \times 6 \times 6$, and $15 \times 5 \times 5$ (number of jobs × number of machines × number of operations), respectively.

TABLE 5.2

Modified instance "mk1" by Brandimarte (1993)

J	O_{jo}	M_1	M_2	M_3	M_4	M_5	M_6	Mc_j
J_1	O_{11}	5	–	4	–	–	–	120
	O_{12}	–	1	5	–	3	–	
	O_{13}	–	–	4	–	–	2	
	O_{14}	1	6	–	–	–	5	
	O_{15}	–	–	1	–	–	–	
	O_{16}	–	–	6	3	–	6	
J_2	O_{21}	–	6	–	–	–	–	100
	O_{22}	–	–	1	–	–	–	
	O_{23}	2	–	–	–	–	–	
	O_{24}	–	6	–	6	–	–	
	O_{25}	1	–	–	–	–	5	
J_3	O_{31}	–	6	–	–	–	–	65
	O_{32}	–	–	4	–	–	2	
	O_{33}	1	6	–	–	–	5	
	O_{34}	–	6	4	–	–	6	
	O_{35}	1	–	–	–	5	–	
J_4	O_{41}	1	6	–	–	–	5	100
	O_{42}	–	6	–	–	–	–	
	O_{43}	–	–	1	–	–	–	
	O_{44}	–	1	5	–	3	–	
	O_{45}	–	–	4	–	–	2	

(Continued)

TABLE 5.2 (Continued)

Modified instance "mk1" by Brandimarte (1993)

J	O_{jo}	M_1	M_2	M_3	M_4	M_5	M_6	Mc_j
J_5	O_{51}	–	1	5	–	3	–	150
	O_{52}	1	6	–	–	–	5	
	O_{53}	–	6	–	–	–	–	
	O_{54}	5	–	4	–	–	–	
	O_{55}	–	6	–	6	–	–	
	O_{56}	–	6	4	–	–	6	
J_6	O_{61}	–	–	4	–	–	2	85
	O_{62}	2	–	–	–	–	–	
	O_{63}	–	6	4	–	–	–	
	O_{64}	–	6	–	–	–	–	
	O_{65}	1	6	–	–	–	5	
	O_{66}	3	–	–	2	–	–	
J_7	O_{71}	–	–	–	–	–	1	120
	O_{72}	3	–	–	2	–	–	
	O_{73}	–	6	4	–	–	6	
	O_{74}	6	6	–	–	1	–	
	O_{75}	–	–	1	–	–	–	
J_8	O_{81}	–	–	4	–	–	2	100
	O_{82}	–	6	4	–	–	6	
	O_{83}	1	6	–	–	–	5	
	O_{84}	–	6	–	–	–	–	
	O_{85}	–	6	–	6	–	–	
J_9	O_{91}	–	–	–	–	–	1	65
	O_{92}	1	–	–	–	5	–	
	O_{93}	–	–	6	3	–	6	
	O_{94}	2	–	–	–	–	–	
	O_{95}	–	6	4	–	–	6	
	O_{96}	–	6	–	6	–	–	
J_{10}	O_{101}	–	–	4	–	–	2	100
	O_{102}	–	6	4	–	–	6	
	O_{103}	–	1	5	–	3	–	
	O_{104}	–	–	–	–	–	1	
	O_{105}	–	6	–	6	–	–	
	O_{106}	3	–	–	2	–	–	
	pc_k	6	8	7	4	5	6	
	e_k	8	10	7.5	12	9	10.5	

Unspecified operating time implies that the concerned operation is not processed on that machine. The operation times on the machines are considered to be different. For example, in Table 5.2 the 10 by 6 data from Brandimarte instances (Brandimarte, 1993) is shown where ten jobs and six machines indicate that each job has a maximum of 6 operations, and at a time only one job can be processed on one machine. The operational time of different operations for jobs on different machines, the processing cost, and energy consumption of respective machines are also given.

Although various optimisation algorithms are available to solve this problem, we considered a newly developed moth flame optimization algorithm (MFOA) (Mirjalili, 2015) for solving the above-described problem. The above procedure is repeated for all each instance to find the robustness of the proposed algorithm. In Table 5.4, the first column indicates different instances, the second column represents the number of jobs, the respective numbers of machines are represented in column 3, and makespan, processing cost, and energy consumption are represented in columns 4, 5, and 6, respectively.

5.6 Discussion on Experimental Results

The modified MOO-HMFO algorithm helps us to determine what must be the best sequence of the jobs scheduled on the given machines, for a particular job to meet the criterion of the objective function. This helps to maximise profits without any resource or time wastage.

In Table 5.3, the comparative analysis of different fitness values of all instances was carried out under the same algorithm parameters. The results of the best, worst, and average fitness values for each iteration of all the instances are shown in Table 5.3. It also shows the average deviation value, which is a measure of the deviation of best/worst fitness value from the average fitness value. We can observe that after a certain number of iterations, few values become constant; in other words, the termination criteria have been met after a significant number of iterations indicated by constant fitness values. For example, consider the first instance, i.e., mk1. After 500 iterations, the best, worst, and average fitness values are the same indicating that the number of iterations required for getting optimal values is reached. One can observe from the instances that the number of iterations required for obtaining optimum values is different. This is the reason why the time for finding optimum values differs for most of the instances.

Table 5.4 shows the values of makespan, processing cost, and energy consumption for ten different instances. These values represent the minimum values that can be obtained when the inputs were given to each algorithm. For instance, for HMFO, considering mk1 benchmark, we got a makespan value of 113, processing cost value of 700, and energy consumption value of 1093.5. These are the optimum values that can be obtained by the simultaneous optimisation of all three objective functions. To verify the effectiveness of the proposed algorithm, ten different instances are used and their results are shown in Table 5.4. It can be inferred from Table 5.4 that makespan, processing cost, and energy consumption values are independent of the number

TABLE 5.3

The Instances with Different Iterations

Instance	Iteration Number	Best Fitness	Worst Fitness	Average Fitness	Average Deviation
mk1	100	389.5	61.35	225.425	72.78474
	300	985.5	155.4	570.45	72.75835
	500	1,906.5	299.1	1,102.8	72.87813
	700	1,906.5	299.1	1,102.8	72.87813
	1,000	1,906.5	299.1	1,102.8	72.87813
mk2	100	834	131.3	482.65	72.79602
	300	2,153.5	338.85	1,246.175	72.80879
	360	2,836	447	1,641.5	72.76881
	700	2,836	447	1641.5	72.76881
	1,000	2,836	447	1,641.5	72.76881
mk3	100	605.5	96.6	351.05	72.48255
	300	1,689.5	267.1	978.3	72.69754
	500	2,608.5	410.2	1,509.35	72.82274
	1,000	5,323	834.85	3,078.925	72.88502
	1,500	7,223	1,130.7	4,176.85	72.92936
mk5	100	257.5	40.25	148.875	72.96390
	300	515.5	81.05	298.275	72.82709
	500	791	126.2	458.6	72.48147
	810	2,005	316.2	1,160.6	72.75547
	1,000	2,005	316.2	1,160.6	72.75547
mk7	100	2,357	374.1	1,365.55	72.60445
	300	7,320.5	1,166.7	4,243.6	72.50683
	500	10,635.5	1,691.7	6,163.6	72.55338
	700	10,635.5	1,691.7	6,163.6	72.55338
	1,000	10,635.5	1,691.7	6,163.6	72.55338
mt06	100	808.5	127	467.75	72.84874
	216	1,400.5	220.15	810.325	72.83189
	500	1,400.5	220.15	810.325	72.83189
	700	1,400.5	220.15	810.325	72.83189
	1,000	1,400.5	220.15	810.325	72.83189
mt10	100	6,629	1,046.5	3,837.75	72.73142
	300	9,255	1,474.7	5,364.85	72.51181
	500	17,081	2,699.55	9,890.275	72.70501
	700	19,101.5	3,004.4	11,052.95	72.81812
	1,000	32,048	5,031.6	18,539.8	72.86055
mt20	100	9,991	1,572.15	5,781.575	72.80758
	300	28,001.5	4,415.65	16,208.575	72.75732
	500	47,799	7,505.5	27,652.25	72.85754
	700	47,799	7,505.5	27,652.25	72.85754
	1,000	47,799	7,505.5	27,652.25	72.85754

(Continued)

TABLE 5.3 (*Continued*)

The Instances with Different Iterations

Instance	Iteration Number	Best Fitness	Worst Fitness	Average Fitness	Average Deviation
la01	100	15,100.5	2,392.55	8,746.525	72.64571
	250	21,651	3,421.75	12,536.375	72.70543
	500	21,651	3,421.75	12,536.375	72.70543
	700	21,651	3,421.75	12,536.375	72.70543
	1,000	21,651	3,421.75	12,536.375	72.70543
la06	100	9,466	1,480.75	5,473.375	72.94631
	375	41,704	6,656.25	24,180.125	72.47223
	500	41,704	6,656.25	24,180.125	72.47223
	700	41,704	6,656.25	24,180.125	72.47223
	1,000	41,704	6,656.25	24,180.125	72.47223

of jobs and machines, and depend mostly on the type of operations. Taking into consideration not only the number of generation (iteration value) as the deciding factor for the algorithm, but also considering the time complexity of order $O(n^2)$ makes the algorithm more enhanced. From the results, we have realised that the optimisation power of MOO-HMF can project optimal results which help to explore and exploit full capabilities of the machines. This also tells us the sequence in which the machines should be supplied with jobs in order to achieve this optimality, hence making the production quite profitable.

It is evident from Table 5.4 that the solutions of makespan, processing cost, and energy consumption of the proposed algorithm are better than NSGA-II in almost all the instances. With these comparison results, we conclude that the arriving at minimum values with HMFO proves its superiority to determine the optimal values for all three functions.

A Gantt chart has been popularly used to represent a schedule. Figure 5.6 illustrates the maximum completion time for the benchmark instance mt06. The *x*-axis of the Gantt chart denotes processing time and the *y*-axis denotes the machines. While executing the program, the flow model is taken as a complex three-dimensional matrix which covers machines, jobs, and operations. For the values of processing costs and energy units, the input three-dimensional matrix is multiplied by their respective unit values for each instance that we have considered. Then, these newly generated matrices are mutated to respective normalised values so as to ensure dimensionally similar values. The minimum values out of the obtained values are represented row-wise and their sum is computed. This obtained sum is then utilised to form a Gantt chart and analysis was carried out by conducting a Hypothesis Test on the obtained results.

From the graph shown in Figure 5.7, we can clearly differentiate the makespan values obtained from the two different algorithms, HMFO and NSGA-II.

TABLE 5.4

Optimal Values of Makespan, Processing Cost, and Energy Consumption of MOO-HMFO and NSGA-II

Instances	Jobs	Machines	Proposed HMFO			NSGA-II		
			Makespan	Processing Cost	Energy Consumption	Makespan	Processing Cost	Energy Consumption
mk1	10	6	113	700	1,093.5	124	714	1,124
mk2	10	6	172	1,032	1,632	190	1,068	1,658
mk3	15	8	425	2,623	4,185	462	2,654	4,203
mk5	15	4	121	741	1,143	138	762	1,168
mk7	20	5	660	3,983	5,992.5	669	4,013	6,023
mt06	6	6	84	510	806.5	102	533	849
mt10	10	10	2,092	9,340	20,616	3,018	9,402	20,645
mt20	20	5	2,914	16,630	28,255	3,128	16,734	28,288
la01	10	5	1,319	7,986	12,346	1,350	8,034	12,386
la06	15	5	1,239	33,610	6,855	1,298	33,814	6,876

Makespan

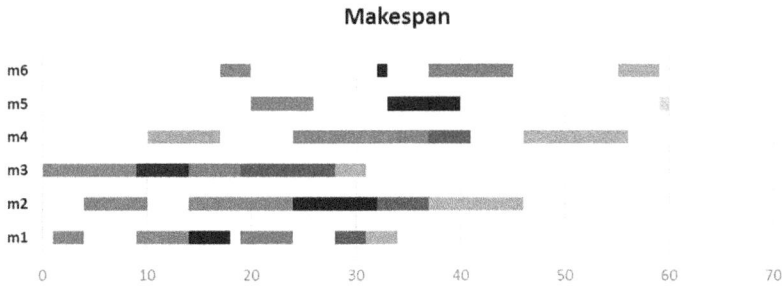

FIGURE 5.6
Gantt chart for the instance mt06.

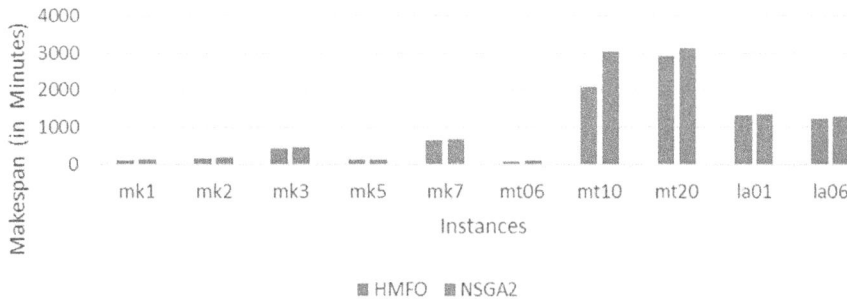

FIGURE 5.7
Comparison study of makespan for all the instances.

For instances with lower makespan, it is observed that the values are approximately the same for both the algorithms indicating that they perform equally well. But as we move towards instances "mt10" and "mt20", a significant difference in makespan values is observed. This difference clearly shows the supremacy of our genetic-based algorithm HMFO over a normal algorithm like NSGA-II. Thus, we can rightfully conclude that our algorithm performs well in both, normal as well as the dynamic environment.

5.7 Conclusions

In this chapter, a multi-objective mathematical model in the context of dynamic FJSSP is developed by considering minimisation of total completion time of jobs, the processing cost of operations, and energy consumption of machines as objective functions subjected to several constraints. To solve the mentioned problem it is necessary to develop an efficient multi-objective evolutionary algorithm due to its complexity and NP-hard nature. Here, the recently introduced and effective evolutionary MFO algorithm is used and the algorithm operators are tuned according to the problem need by proposing it into a MOO-HMF. Moreover, interdependent objectives, i.e., an optimal sequence of jobs, and the number of generations, are also taken into consideration to test the performance of the algorithm along with primary objective functions. The IPPS approaches have been adapted to generate effective process plans which lead to obtaining the optimal/near-optimal solutions. We have tested the objectives and the performance of the algorithm with several instances, and to validate the results a comparison study has been conducted with the benchmark algorithm, i.e., NSGA-II.

References

Brandimarte, P. (1993). Routing and scheduling in a flexible job shop by tabu search. *Annals of Operations Research*, 41(3), pp. 157–183.

Deb, K., Pratap, A., Agarwal, S. and Meyarivan, T.A.M.T., (2002a). A fast and elitist multiobjective genetic algorithm: NSGA-II. *IEEE Transactions on Evolutionary Computation*, 6(2), pp. 182–197.

Fang, H.L., Ross, P. and Corne, D., (1993). *A Promising Genetic Algorithm Approach to Job-Shop Scheduling, Rescheduling, and Open-Shop Scheduling Problems* (pp. 375–382). University of Edinburgh, Department of Artificial Intelligence, Berlin, Heidelberg.

Fang, K., Uhan, N., Zhao, F. and Sutherland, J.W. (2011). A new approach to scheduling in manufacturing for power consumption and carbon footprint reduction. *Journal of Manufacturing Systems*, 30(4), pp. 234–240.

Fisher, H. and Thompson, G.L. (1963). Probabilistic learning combinations of local job-shop scheduling rules, J.F. Muth and G.L. Thompson (Eds.), *Industrial Scheduling*. Prentice-Hall, Englewood Cliffs, NJ, pp. 225–251.

Gao, J., Gen, M., Sun, L. and Zhao, X. (2007). A hybrid of genetic algorithm and bottleneck shifting for multiobjective flexible job shop scheduling problems. *Computers and Industrial Engineering*, 53(1), pp. 149–162.

Gen, M. and Lin, L. (2007). Genetic algorithms. *Wiley Encyclopedia of Computer Science and Engineering*, pp. 1–15.

Gen, M., Gao, J. and Lin, L. (2009). Multistage-based genetic algorithm for flexible job-shop scheduling problem. *Intelligent and Evolutionary Systems*, pp. 183–196.

Gonçalves, J.F., de Magalhães Mendes, J.J. and Resende, M.G. (2005). A hybrid genetic algorithm for the job shop scheduling problem. *European Journal of Operational Research*, 167(1), pp. 77–95.

Jiang, Z., Zuo, L. and Mingcheng, E. (2014). Study on multi-objective flexible job-shop scheduling problem considering energy consumption. *Journal of Industrial Engineering and Management*, 7(3), p. 589.

Karthikeyan, S., Asokan, P., Nickolas, S. and Page, T. (2015). A hybrid discrete firefly algorithm for solving multi-objective flexible job shop scheduling problems. *International Journal of Bio-Inspired Computation*, 7(6), pp. 386–401.

Li, J.Q., Pan, Q.K. and Gao, K.Z. (2011). Pareto-based discrete artificial bee colony algorithm for multi-objective flexible job shop scheduling problems. *The International Journal of Advanced Manufacturing Technology*, 55(9), pp. 1159–1169.

Liu, Y., Dong, H., Lohse, N., Petrovic, S. and Gindy, N. (2014). An investigation into minimizing total energy consumption and total weighted tardiness in job shops. *Journal of Cleaner Production*, 65, pp. 87–96.

Madureira, A. and Pereira, I. (2010, August). Intelligent bio-inspired system for manufacturing scheduling under uncertainties. *In Hybrid intelligent systems (HIS), 2010 10th international conference* on (pp. 109–112). IEEE, Atlanta, GA.

Madureira, A., Pereira, I., Pereira, P. and Abraham, A. (2014). Negotiation mechanism for self-organized scheduling system with collective intelligence. *Neurocomputing*, 132, pp. 97–110.

May, G., Stahl, B., Taisch, M. and Prabhu, V. (2015). Multi-objective genetic algorithm for energy-efficient job shop scheduling. *International Journal of Production Research*, 53(23), pp. 7071–7089.

Mirjalili, S. (2015). Moth-flame optimization algorithm: A novel nature-inspired heuristic paradigm. *Knowledge-Based Systems*, 89, pp. 228–249.

Pezzella, F. and Merelli, E. (2000). A tabu search method guided by shifting bottleneck for the job shop scheduling problem. *European Journal of Operational Research*, 120(2), pp. 297–310.

Reddy, M.S., Ratnam, C., Agrawal, R., Varela, M.L.R., Sharma, I. and Manupati, V.K. (2017). Investigation of reconfiguration effect on makespan with social network method for flexible job shop scheduling problem. *Computers and Industrial Engineering*, 110, 231–241. doi:10.1016/j.cie.2017.06.014.

Santos, A.S., Madureira, A.M., Varela, M.L.R., Putnik, G.D., Kays, H.E. and Karim, A.N.M. (2015, June). Scheduling and batching in multi-site flexible flow shop environments. *In Information Systems and Technologies (CISTI), 2015 10th Iberian Conference on Information Systems and Technologies (CISTI 2015)*, pp. 1–6), IEEE/ IEEE Xplore, Agueda, Aveiro, Portugal. doi:10.1109/CISTI.2015.7170525.

Silva, C., Reis, V., Morais, A., Brilenkov, I., Vaza, J., Pinheiro, T.,Neves, M., Henriques, M., Varela, M.L., Pereira, G., Fernandes, N.O., Carmo-Silva, S. and Dias, L. (2017). A comparison of production control systems in a flexible flow shop. *Procedia Manufacturing*, 13, pp. 1090–1095. doi:10.1016/j.promfg.2017.09.169.

Tang, D., Dai, M., Salido, M.A. and Giret, A. (2016). Energy-efficient dynamic scheduling for a flexible flow shop using an improved particle swarm optimization. *Computers in Industry*, 81, pp. 82–95.

Varela, M.L., Trojanowska, J., Carmo-Silva, S., Costa, N.M. and Machado, J. (2017). Comparative simulation study of production scheduling in the hybrid and the parallel flow. *Management and Production Engineering Review*, 8(2), pp. 69–80. doi:10.1515/mper-2017-0019.

Wolpert, D.H. and Macready, W.G., (1997). No free lunch theorems for optimization. *IEEE tTransactions on eEvolutionary cComputation*, 1(1), pp. 67–82.

Yagmahan, B. and Yenisey, M.M., (2008). Ant colony optimization for multi-objective flow shop scheduling problem. *Computers &and Industrial Engineering*, 54(3), pp.411–420.

Yin, L., Li, X., Gao, L., Lu, C. and Zhang, Z., (2017). Energy-efficient job shop scheduling problem with variable spindle speed using a novel multi-objective algorithm. *Advances in Mechanical Engineering*, 9(4), doi:10.1177/1687814017695959.

Zhang, G., Shao, X., Li, P. and Gao, L. (2009). An effective hybrid particle swarm optimization algorithm for multi-objective flexible job-shop scheduling problem. *Computers and Industrial Engineering*, 56(4), pp. 1309–1318.

Zhang, R. and Chiong, R. (2016). Solving the energy-efficient job shop scheduling problem: a multi-objective genetic algorithm with enhanced local search for minimizing the total weighted tardiness and total energy consumption. *Journal of Cleaner Production*, 112, pp. 3361–3375.

6

Integration of Scheduling and Process Planning in Shop Floor: A Probability Model-Based Approach

R. Pérez-Rodríguez
CONACYT - Center for Mathematics Research México

A. Hernández-Aguirre
Center for Mathematics Research México

S. Frausto-Hernández
Aguascalientes Institute of Technology

CONTENTS

6.1 Introduction 135
6.2 Problem Statement 136
6.3 Probability Models 137
    6.3.1 A Probability Model for the Process Plan Selection 137
    6.3.2 A Probability Model for the Operation Scheduling Decision 138
6.4 Results and Comparison 138
6.5 Conclusion 139
References 140

6.1 Introduction

The job shop environment has been studied extensively for many decades. Wide and diverse algorithms with different characteristics have been proposed. The scheduling and process planning integration, for the job shop floor, is an interesting topic to enhance the throughput of any manufacturing system. An extensive review of this topic has been presented in Chapter 2. Most approaches, for solving the job shop environment, consider the scheduling and process planning as independent events. The goal of this research is to model interactions between scheduling and process planning. Specifically, the interactions can be estimated through the key variables in the process planning and scheduling functions. Some algorithms try to estimate the dependencies and correlations among the key variables of the problem. In some situations, the range of values for each key variable is not affected by the range of values of the rest of the variables, i.e., a relationship does not exist. In other circumstances, the range of values for each key variable may be affected by the range of values of another variable, i.e., a bivariate relationship exists. Lastly, the range of values for each key variable may be affected by the range of values of more than two variables, i.e., a diverse relationship exists. The key variables, for process planning, can be the alternative process plans of a job. The key variables, for scheduling, can be the sequences of jobs over alternative machines. All the variables are represented in solution vectors. Each solution vector contains different values for each variable. Therefore, a solution is detailed by all the elements contained in a solution vector.

In this study, *Estimation of Distribution Algorithms* (EDA) from Mühlenbein and Paaß (1996), is used to identify the interactions between the variables of the problem. Basically, a probability model is built using statistical information in order to build an EDA. The information is depicted in all the elements guarded in the solutions.

6.2 Problem Statement

The integration of process planning and scheduling (IPPS) in a job shop floor is detailed as follows: there are n jobs $J = \{J_1, J_2, \ldots, J_n\}$ to be processed on m machines $M = \{M_1, M_2, \ldots, M_m\}$. A job J_i is formed by a sequence of n_i operations $\{O_{i,1}; O_{i,2}; O; \ldots; O_{i,n_i}\}$ performed one after another according to a given sequence, i.e., the corresponding operations have to be set in motion in the given arrangement. Then, for an operation, the starting time must not be earlier than the point at which the anterior operation is completed. In addition, each operation must be completed without interruption once it starts. The execution of $O_{i,j}$ requires one machine out of a set of $m_{i,j}$ given machines $M_{i,j} \subseteq M$, it is imposed for all applicable pairs of precedence operations. The objective is to minimise the makespan, i.e., the total completion time for all the current orders. The IPPS considers multiple plans for each job by excluding the conjecture of only one plan being available. Consider a group of n jobs, each having a group of P plans. Then, any plan p_k of the job $j(p_{kj})$ is an arranged index of n_i operations. The feasible plans can be "admitted" to "in progress". The operation i of a job j in the plan $p_{kj}(O_{i,j,p_{kj}})$ desires one machine out of a class of $m_{i,j}$ given machines $M_{i,j} \subseteq M$. Finally, there should be one adopted plan for every job.

6.3 Probability Models

Any solution to the integration of scheduling and process planning should be a combination of machine assignment, and process plan selection. In other words, the selection of the assignment of operations on machines, and a plan for each job should express a solution. Therefore, two different solution vectors can be identified in any solution. In addition, each vector type can generate a probability model. The easiest way to build a probability model is without interactions, i.e., the range of values for each variable are not affected by the values of the remaining variables.

6.3.1 A Probability Model for the Process Plan Selection

In a process plan vector, each element shows a variable. Each position in the vector indicates which process plan is selected for each job. An illustration is depicted in Figure 6.1 with six jobs. Each job has three possible process plans to choose from.

As in any EDA, in this study also, we compute different vectors. All these vectors form a population. It is named N. Then, with these vectors, we can build a probability model. We use the *Univariate Marginal Distribution Algorithm* (UMDA) detailed by Mühlenbein (1997). It estimates a combined probability distribution of the chosen solution vectors in every iteration, $p(x)$. The combined probability distribution can be expressed as a consequence of univariate marginal distributions, $p(x) = p(x|N) = \prod_{i=1}^{n} p(x_i)$.

Each univariate marginal distribution is computed from marginal frequencies, i.e., from solution vectors $p(x_i) = \dfrac{\sum_{j=1}^{N} \delta_j (X_i = x_i | N)}{N}$

where

$$\delta_j (X_i = x_i | N) = \begin{cases} 1, & \textit{if the jth case of } N, X_i = x_i \\ 0, \textit{otherwise} \end{cases}$$

Figure 6.2 depicts how to compute the probability distribution on a population N of process plan vectors. The value $p_{kj}(l)$ expresses the probability that a plan k is chosen for the job j at generation l. The value of p_{kj} shows the significance of a plan for a determined job for all $k(k = 1,2,...,P_j)$ and $j(j = 1,2,...,n)$.

	Process plan vector					
1	3	3	1	2	2
Job 1	2	3	4	5	6

FIGURE 6.1
Solution representation of a process plan vector.

Process plan vectors

Job   1 2 3 ... ...

Job 1

$$
\begin{bmatrix}
1 & 1 & 3 & 4 & 4 & 2 \\
2 & 1 & 4 & 1 & 3 & 1 \\
2 & 2 & 2 & 3 & 2 & 1 \\
2 & 2 & 3 & 4 & 3 & 2 \\
2 & 3 & 2 & 1 & 2 & 1 \\
2 & 2 & 2 & 1 & 2 & 2 \\
2 & 2 & 3 & 1 & 4 & 2 \\
1 & 1 & 1 & 1 & 2 & 2 \\
1 & 2 & 1 & 3 & 1 & 1 \\
1 & 3 & 2 & 1 & 3 & 2
\end{bmatrix}
$$

p(X1=1|N) = 4/10
p(X1=2|N) = 6/10

Job 2
p(X2=1|N) = 3/10
p(X2=2|N) = 5/10
p(X2=3|N) = 2/10

N Selected population

B1 matrix   Job

	1	2	3	...	...
1	0.4	0.3	...		
2	0.6	0.5	...		
3	....	0.2	...		
...	....	....	....		
...					

Process plan

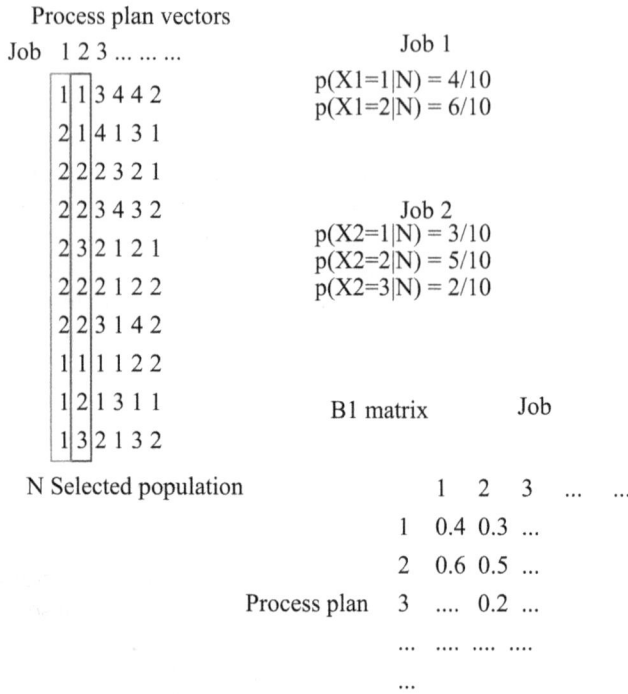

FIGURE 6.2
A probability model for the process plan decision.

6.3.2 A Probability Model for the Operation Scheduling Decision

In the machine assignment vector, each element depicts the selected machine for each operation. To obtain the probability model we use the UMDA algorithm, as described in the previous step. The element $p_{ijm}(l)$ of the probability-model, expresses the probability that operation $O_{i,j}$ is executed on a machine m at generation l. Then, p_{ijm} shows the rationality of an operation executed on a machine.

6.4 Results and Comparison

The relevance of this paper is validated by a comparison of the results with other algorithms on certain general and standard benchmarking datasets. We measure the *Relative Percentage Increase* (RPI).

$$
RPI(c_i) = \frac{\left(c_i - c^*\right)}{c^*}
$$

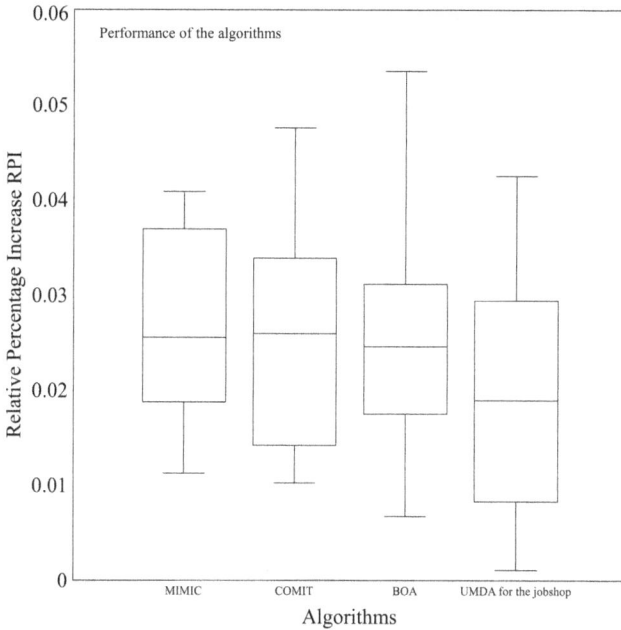

FIGURE 6.3
Performance of the algorithms.

where c_i depicts the overall completion time for all the jobs obtained in the i-th replication and c^* is the best total completion time for all the jobs found and reported in the literature.

Some algorithms are used in this study. *Mutual Information Maximising Input Clustering* (MIMIC) algorithm proposed by De Bonet et al. (1997), *Combining Optimisers with Mutual Information Trees* (COMIT) algorithm based on Baluja and Davies (1997), and *Bayesian Optimisation Algorithm* (BOA) based on Pelikan et al. (1998) are used as a reference for contrast with the proposed design. Figure 6.3 details the performance of the schemes used for the comparison. It can be easily observed that the proposed probability model outperforms all the other algorithms.

6.5 Conclusion

This chapter discusses the integration between scheduling and process planning in a job shop environment. The solution representation used in this study is suitable to implement probability models. The proposed algorithm offers a better estimation of each operation in the machine assignment vector for the job shop environment. Other probability distributions for the integration between scheduling and process planning can be studied in prospect. Computing interactions for other dynamic job shop concerns, i.e., shutdown, maintenance, different work orders, online systems, can be studied in light of these results.

References

Baluja, S. & Davies, S. (1997). Using optimal dependency-trees for combinatorial optimization: learning the structure of the search space. In D. Fisher (Ed.), *Proceedings of the Fourteenth International Conference on Machine Learning.* San Francisco, CA: Morgan Kaufmann Publishers Inc., pp. 30–38.

De Bonet, J., Isbell, C. & Viola, P. (1997). MIMIC: finding optima by estimation probability densities. *Advances in Neural Information Processing Systems,* 424–430.

De Bonet, J. S., Isbell Jr, C. L. & Viola, P. (1997). MIMIC: Finding optima by estimating probability densities. *Advances in Neural Information Processing Systems 9: Proceedings of the 1996 Conference* (Vol. 9, pp. 424–430). MIT Press.

Mühlenbein, H. & Paaß, G. (1996). From recombination of genes to the estimation of distributions. I binary parameters. *Proceedings of the 4th International Conference on Parallel Problem Solving from Nature* (Vol. 187). Berlin: Springer.

Mühlenbein, H. (1997). The equation for response to selection and its use for prediction. *Evolutionary Computation,* 5(3), 303–346.

Pelikan, M., Goldberg, D. E. & Cantu-Paz, E. (2000). Linkage problem, distribution estimation, and Bayesian networks. *Evolutionary Computation,* 8(3), 311–340.

7

Integrated Process Planning and Scheduling Using Dynamic Approach

Gideon Halevi

Retired – Director of CAD/CAM R&D Center at IMI Corporation

Rakesh Kumar Phanden

Amity University Uttar Pradesh

CONTENTS

7.1   Introduction 141
7.2   Gaps Between Process Planning and Scheduling 142
      7.2.1   Example—Process Planning 143
            7.2.1.1   Solution 144
      7.2.2   Example—Decision on Sequence of Operations 144
7.3   IPPS 146
7.4   Process Planning 147
      7.4.1   Practical Process—Operation Transformation 148
7.5   Scheduling 150
      7.5.1   Concept and Terminology 151
7.6   IPPS—Algorithm and Terminology 152
      7.6.1   Example 154
            7.6.1.1   Economic Calculations 155
            7.6.1.2   Possibility of Disruption 157
7.7   Conclusion 157
7.8   Appendix 158
References 159

7.1 Introduction

Process planning is a very flexible technical task. It may generate several processes that produce an item as detailed by the engineering drawing. Scheduling, on the contrary, is a mathematical task, searching for optimum sequencing operations on resources. The transfer of data from process planning to scheduling is by a routine, i.e., a set of specific operations required to produce an item. This put a constraint on the scheduler and makes scheduling a complex task. Integration of process planning with scheduling (IPPS) increases tremendously the number of scheduling options, i.e., routine is variable, the sequence of operations is unpredictable, and resource selection is random (Jain et al. 2006; Phanden et al. 2011, 2013). Producing a variety of products at a minimal cost and within a specified time is the main objective of production scheduling function. Thus, scheduling becomes even more complex. The proposed IPPS approach is based on the scheduling rules. The lead time and processing cost have been reduced by using the methodology of integrated dynamic planning (as discussed in the example in Section 7.2.2, the lead time is reduced to 21 units from 35 units and processing cost is reduced to 101 units from 162 units). The proposed plan is realistic and also avoids the possibility of multiple disruptions such as machine breaks, bottlenecks, newly arrived orders, etc. on the shop floor. The mathematical research power within scheduling may be directed towards generating economical rules, such as, selecting not the best process, but the one that results in increasing company profit. One method to achieve IPPS has been presented in this chapter.

7.2 Gaps Between Process Planning and Scheduling

Process planning has always been an element of paramount importance in manufacturing as it determines how any part is to be manufactured. It plays a key role in determining the cost of parts and affects various shop floor activities, production planning function, quality of the product, production rate as well as the company's affordability to produce. Also, it acts as a bridge between designing and manufacturing (Halevi and Kals, 1997; Halevi and Wang, 2007).

Though process planning is highly important for manufacturing, there exists no proper method that proves to be useful, or assists in training people for such an important task. It involves a huge amount of labour and completely depends upon experiences, understanding, and skill of the process planner. Thus, frequently it lacks in-depth optimisation and proper analysis of process plans and often results in more than the specified cost of production, untuned processes, and unnecessary delays (Halevi, 2012).

It has been observed that *process planning is an art, not a science*. Various studies clearly reveal that different experts have expressed their own knowledge and experiences; one expert's practice significantly differs from another. Therefore, the chances are very rare that two planners produce similar process plan for one product in the same manufacturing setting. Specifications proposed in engineering drawings are used to recommend the process and it produces the part accordingly. However, various recommended processes yield diverse processing times along with different production costs.

Here, the question arises *who is an expert?* An expert can be defined as someone who is experienced, skilled, and well-informed to attain success in a specific domain, or one who has attained special skill and mastery in doing something. Even so, experience is something which is gained during the actual work throughout the production, when defined procedures are laid down, along with regular follow-ups and remedial measures undertaken to accomplish the job. Experience is acquired from the problematic and complex processes, such as from the rejection of parts which are discarded and from the improvements made in order to gain fruitful solutions. Even the possibility of gaining a small or an erroneous experience exists, if process plans possessed *no problem* kind of processes.

It is worth sharing that after working with more than around 250 experienced process planners and analyzing their way of working, I (Gideon Halevi) can conclude that experience gained is priceless if taken rightly, although generally that's not the case. Apparently, decisions were taken by process planners according to their overall understanding instead of segregating the problem into individual parameters. In fact, they did understand the problem and determined a feasible solution. However, the controlling factors which triggered the problem are beyond their level of comprehension. Consequently, even though controlling factors differ from each other, the same solution method is applied to various identical problems by process planners. Process planners are proficient to handle processes, it is not their job to make economic decisions of a company. Thus, they should be responsible to provide the data to other respective departments of the company and let each department make its own decision in order to select optimum process plans.

Process planner may create various alternative methods to process a job, thus he is flexible in his decision making. Although, he does not have the parameters required to make other reasonable selections. The data needed for other realistic decisions (to choose alternative process plans) are available with the scheduler, such as the production-order quantity demand, availability of resources at the shop floor, the criterion for optimisation, identification of critical jobs and processes, maximum production, and/or minimum cost criterion, etc. However, the scheduler communicates with the process planner only to ask for the generated operation route. Thus, the process planner has no communication with his follower (scheduler). Therefore, generating a production schedule with one route constrains the scheduler and makes his chore very complex. Many scheduling problems can be solved by altering the process route (Halevi, 1997, 2017).

7.2.1 Example—Process Planning

A part-drawing is demanded a plate having hole and pockets. It needs to remove the raw material of 6 mm by the tolerance of 0.1 mm. In order to achieve these requirements, it must follow two metal cutting passes, i.e., one rough cut and one finish cut as shown in Table 7.1.

TABLE 7.1

Specifications of Operations

Operations No.	Machine's Power Consumption (kW)	Processing Time (min)
001 – Rough cut	19	0.60
002 – Finish cut	3	0.94
003 – Rough pocket	1.6	0.90
004 – Finish pocket	0.2	4.86
		Total = 15.40

7.2.1.1 Solution

(i) Machine having power consumption of 19 kW or more with 15.4 min for each item: the scheduler should follow the routing procedure in which the resource selected, for operation 001, is consuming (15.4 – 0.60) 14.8 min under loading conditions but restrictions are imposed on this machine for those jobs which require a high amount of power. The process planner, after reducing the cutting speed designed for rough operation, might also propose a plan to select a 3 kW machine. Hence the processing time is increased from 0.6 to 3.8 min, thereby the new routing procedure will be (ii) machine having a power consumption of 3 kW for a period of 18.6 min: therefore, a machine with a high-power rating can be used by other parts as the operation time has been increased by 3.2 min.

There is a possibility that process planner does not have any specific intention to choose such routing procedure. But it may solve several bottlenecks in shop floor scheduling. If the production planning is aware of such possible changes it will add scheduling options.

7.2.2 Example—Decision on Sequence of Operations

The planner of processes must make a decision even if he does not possess any preference or technical priorities. As discussed in Halevi (1980), the Figure 7.1 shows eight operations which are needed to make a certain part.

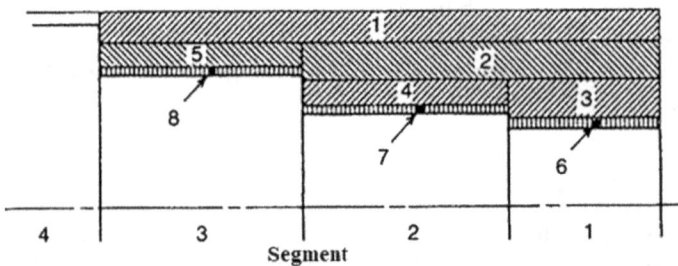

FIGURE 7.1
Specified sequence of operations.

Unless operation 1 is carried out, operation 2 cannot be performed. Similarly, operations 6 and 3 have to be conducted in that specific order. Although the second operation makes a path for the tool which processes the fifth procedure, hence the fifth procedure might be processed after the second procedure. The eighth procedure might be processed after the fifth procedure and the seventh procedure after the fourth operation.

Thus, the allowable sequence of operations might be: *1-2-3-6-4-5-7-8* or *1-2-5-8-3-4-6-7* or *1-2-3-5-8-4-7-6* or *1-2-3-4-6-7-5-8* or *1-2-5-3-4-6-7-8* or *1-2-5-8-3-6-4-7* or *1-2-3-4-5-8-7-6*.

There is top notch flexibility to the planning of the process, but the method incorporated prohibits the planner of the process to transfer it to the scheduler, and for it, to get benefit from management. To validate the feasibility of dynamic planning method and to check the time of execution, a program of computers has been constructed (Halevi, 1998, 2000). The program of testing assumes a firm that makes pneumatic cylinders of two kinds: double acting and spring return. The example data considered two orders, 120 spring return cylinders are to be taken on the 40th day and 40 double-acting cylinders are to be taken on the 35th day. The total data include 12 parts, 30 operations, and 15 resources. A matrix of the routine procedure is built for each operation. The sequence of the operations is variable, and each operation may be processed on any of the 15 resources. The scheduling is decided by an algorithm for best results.

The same set was run with several optimisation criteria such as mix mode, dynamic planning method. It took about 20 s to elapse time to get the required schedule. The outcomes are shown in Table 7.2 (Halevi, 1999).

This example proves that the dynamic planning method is favourable due to time and cost parameters. Moreover, it proves that the plan that is released is empirical, and therefore might remove multiple disturbances on the floor of the store that occur because of system rigidity. It also proves that integration will make scheduling simpler. It will improve productivity, introduce costing, reduce throughput time, etc. One method to achieve IPPS is presented in this chapter.

TABLE 7.2

Results Obtained with Respect to the Different Optimisation Criterion

Optimisation Criterion	No. of Periods	Units of Cost
Minimum cost	32	76.2
Maximum production	35[a]	162
Mixed cost and time	23	131
Flexible algorithm	21	101

[a] It is not a mistake since most operations selected the best machine, it was overloaded, and the cost raised accordingly. The flexible method uses more machines as shown in Table 7.3.

TABLE 7.3

Machines Used in a Flexible Method

Periods	Machines

Periods	1	2	3	4	5	6	7	8	9	10	11	12	13	14	15
1	—	—	—	—	—	—	—	1,008	903	702	401	—	—	—	—
2	—	—	—	—	—	—	—	903	907	702	401	—	—	—	—
3	—	—	—	—	—	—	—	907	803	702	401	—	—	—	—
4	—	—	—	—	—	—	—	1,004	—	702	401	—	—	—	—
5	301	—	—	—	602	—	—	1,004	—	706	405	—	—	—	—
6	301	—	305	—	602	1,206	—	807	—	—	—	—	—	—	—
7	301	—	305	—	602	506	—	—	—	—	—	—	—	—	—
8	301	—	602	—	—	506	—	—	—	—	—	—	—	—	—
9	301	—	205	602	—	—	—	—	—	—	—	—	—	—	—
10	301	—	205	—	502	—	—	—	—	—	—	—	—	—	—
11	—	201	—	—	502	—	—	—	—	—	—	—	1,105	—	—
12	—	201	—	—	502	—	—	—	—	—	—	—	—	—	—
13	—	201	502	—	—	—	—	—	—	—	—	—	—	—	—
14	—	201	502	—	—	—	—	—	—	—	—	—	—	—	—
15	—	201	502	—	—	—	—	—	—	—	—	—	—	—	—
16	—	201	502	—	—	—	—	—	—	—	—	—	—	—	—
17	—	201	502	—	—	—	—	—	—	—	—	—	—	—	—
18	—	—	—	—	—	—	—	—	—	—	—	—	101	—	—
19	—	—	—	—	—	—	—	—	—	—	—	—	101	—	—
20	—	—	—	—	—	—	—	—	—	—	—	—	101	—	—
21	—	—	—	—	—	—	—	—	—	—	—	—	101	—	—
22	—	—	—	—	—	—	—	—	—	—	—	—	—	—	—

I (Gideon Halevi) have spent 2 years of my life attempting to generate a route-process for a matrix of 10 × 10 operation-resource with the option of changing sequence. I have used improved Bellman theory of dynamic programming. It was not quite a total waste as I got my PhD for it. Finally, I realised that instead of presenting the scheduler with one route, I may present the scheduler the full matrix with infinite process options, let him take the one that makes the scheduling decision. It possesses all the knowledge and information to do the decision. This is called IPPS in practice.

7.3 IPPS

The purpose of IPPS is to increase plant productivity and to reduce the manufacture cost. The considered method is by setting innovative schedule options such as a routine becomes variable, the sequence of operations is changeable, and all resources are optional. Scheduling objective is to produce a given product mix at a specified time at minimal cost. Thus, IPPS will enable the process planner flexibility (Jain et al., 2013), he may recommend a set of process options and let the scheduling freedom to select the one that will serve its objections best, furthermore, it will introduce a cost of processing as a parameter of scheduling. Moreover, it enables to follow order status from order to delivery.

7.4 Process Planning

The process planning system is made of two stages: transformation and technology. The stage of technology makes a Theoretical Process (TP). The stage of transformation adjusts the TP to become practical for a specific plant resource and specific order quantity. The matrix does not present a routine, however, it presents several options for scheduling, leaving the decision of which process to select to the scheduling stage (will be detailed in the following section). Not making decisions at this stage eliminate introducing artificial constraints.

The definition of TP is *the best process that is possible from the viewpoint of technology, i.e., it does not breach any physical or technological rule.* It is conceptual from the point of view of a shop, i.e., the shop that is specific might not have the resources that are required, or that the resources that are required are not market available. It is theoretical because *resources that are imaginary* are defined and the CAPP system or process planner, will make a process considering these resources and ignoring all elements that are variable such as handling time, order quantity, set-up time, etc. (Halevi and Roland, 2012).

The imaginary resource capabilities are according to the boundaries that are present because of technology. The strength is in regard to the largest motor of electric nature that could be constructed at any place in the globe which signifies no force or power constraints. The tools that are imaginary are constraints of the present day technology. This means—usage of the best grade tool, the toughest and hardest possible. The rigidity can be ignored and is absolute.

However, the cutting forces (comprising of feed rate and depth of cut) are constraint by the strain and strength of the part. The cut depth is constraint by the tolerances. The rate of feed is constraint by the finish on the surface; the chucking is constraint by the shape and size of the parts.

The output of this stage is the TP time of each operation, the relationship and priority constraints and parameters that were incorporated to compute and specify the operations of theoretical nature. The data are specific for every processing type and will be incorporated in the stage of transformation. It should include all the parameters that are needed for the transformation

TABLE 7.4

Process Operation Matrix

Operation No.	Operation Type	Priority	TD	L	D	F	S	P	T
001	RMO	0	125	378	4.4	808	100	20	0.47
002	RMO	1	125	128	4.6	735	100	20	0.17
003	SFMO	2	125	278	0.4	905	148	2.2	0.31
004	FMO	3	125	378	0.2	200	165	0.39	1.89
005	RPMO	1	80	150	4.0	1,093	102	20.6	0.24
006	FPMO	5	12	472	0.4	120	24	0.33	4.16
007	CDO	2	3	3	—	0.05	14	0.025	0.03
008	TDO	7	7	21	—	0.16	15.7	0.3	0.22
009	CDO	8	12	21	—	0.19	23.5	0.5	0.20

TD, Tool Diameter, L, Length, D, Depth, F, Feed in mm per min, S, Speed in m per min, P, Power in kW, T, Time in min, RM, Rough Milling Operation, SFMO, Semi Finish Milling Operation, FMO, Finish Milling Operation, RPMO – Rough Pocket Milling Operation, FPMO, Finish Pocket Milling Operation, CDO, Center Drilling Operation, TDO, Twist Drilling Operation, CDO, Core Drilling Operation.

(Halevi, 1999). Table 7.4 demonstrates the matrix format for a part "plate". The tool diameter, length, and depth are in millimeter. The process planner calls for nine operations. Their name, required power, cutting conditions, and time of machining are shown in the proper column.

The priority column indicates the restrictions on the machining operation sequence. The 0 priority suggests that the functioning might be performed anywhere or be the first operation of machining. Usually, the constraint is because of access to tools, as can be seen, operation 010 can be the operation that commences. After operation 001 is when operation 002 can be made. Operation 006 could be made after operation 005. In the column, the number specifies that the functioning, that is relevant, can be linked after the number indicating operations in the priority column are done. Although, the operation does not have to be followed immediately. Thereby, the operational sequence may be: 001; 009; 008; 007; 006; 005; 004; 003; 002; *OR* 001; 003; 002; 008; 007; 009; 005; 004; 006; *OR* 001; 002; 005; 003; 007; 004; 008; 006; 009; *OR* numerous more combinations. Other constraints on operation sequences such as geometric tolerances can be indicated by codes in the priority or relation column. Thus, the matrix indicates several alternatives from which scheduling options need to be selected.

7.4.1 Practical Process—Operation Transformation

To consider the practical process the TP is adjusted and translated to the constraints and capabilities which comply with the specific machine. It is clear that the time of machining cannot be reduced, just boosted. The modification takes the following factors into consideration: special features, machine physical size, the accuracy of the machine, available torque & power, available feeds and speeds, handling time, tools number, controls type, etc. Times of direct handling, like tool change, speed of adjustment, etc., are added to the time of operation. In the event that a machine cannot perform an operation because of accuracy or more reasons, it stays inside the matrix, and its time of machining is marked as high values (99). This prevents the machine's selection for that operation but can also be selected for different operations. Inaccurate, low cost, old machines might be incorporated for functionalities that are rough (Halevi, 1999).

This feature of matrix construction enables economical evaluation of incorporating machines that are conventional versus CNC machines, operations that are manual versus operations that are assisted by tools, etc. Example of practical process plan in a matrix format for six machines is shown in Table 7.5. Thus, the matrix contains Tij values, where Tij accounts for the machine to time operation (i) on the machine (j).

The conversation from theoretical to the practical process may be done manually by process planner or by a computer flow chart program that includes the required equations. The content of the time matrix can be changed easily to a matrix of cost. This is performed by increasing the time rate hourly. The hourly rate of each machine can be given as company rate or as a relative number to be set by company management. Thus, the value Tij is converted to Cij, where Cij represents the operation price of performance i on the j machine. The values of the time that are changed to values of cost nature are presented in Table 7.6. These two matrices are the base of the proposed scheduling method. The term BEST is different between scheduling (mathematics) and process planning. In Table 7.5 "Total" on R1 is 8.59 and on R2 is 9.04. Now check in Table 7.6. Total is 34.36 and on R2, it is 27.12; who is the best?

TABLE 7.5

Resource – Operation Time Matrix

				R	E	S	OU	R	CE
Operation	Time (in min)	Priority	Rel	#1	#2	#3	#4	#5	#6
001	0.47	0	0	0.57	0.62	1.28	99	1.62	1.19
002	0.17	1	0	0.27	0.32	0.88	99	1.22	0.59
003	0.31	2	0	0.41	0.46	0.97	99	99	0.56
004	1.89	3	0	1.99	204	2.55	99	99	2.14
005	0.24	1	0	0.34	0.39	0.99	99	1.32	0.74
006	4.16	5	0	4.26	4.31	4.38	99	99	4.41
007	0.03	2	0	0.13	0.18	0.69	0.69	1.03	0.28
008	0.22	7	0	0.32	0.37	0.88	0.88	1.22	0.47
009	0.20	8	0	0.30	0.35	0.86	0.86	99	0.45
Total	7.69	—	—	8.59	9.04	13.92	—	—	10.82

TABLE 7.6

Resource – Operation Cost Matrix

Operation	Time (in min)	Priority	Rel	R #1	E #2	S #3	OU #4	R #5	CE #6	Minimum Cost
001	0.47	0	0	2.28	1.86	1.79	99	1.62	2.36	1.62 M5
002	0.17	1	0	1.08	0.96	1. 23	99	1.22	1.22	0.96 M2
003	0.31	2	0	1.64	1.38	1.36	99	99	1.12	1.12 M6
004	1.89	3	0	7.96	6.12	3.57	99	99	4.28	3.57 M3
005	0.24	1	0	1.36	1.17	1.39	99	1.32	1.48	1.17 M2
006	4.16	5	0	17.04	12.93	6.75	99	99	8.82	6.75 M3
007	0.03	2	0	0.52	0.54	0.97	0.69	1.03	0.56	0.52 M1
008	0.22	7	0	1.28	1.1	1.24	0.88	1.22	0.94	0.88 M4
009	0.20	8	0	1.20	1.05	1.20	0.86	99	0.90	0.86 M4
Total	7.69	—	—	34.36	27.12	19.50	—	—	21.64	17.45

For low quantity, select the best process using the Total value (pay once for transfer penalty). For high quantity, select for each resource the best operation. In between, check the differences and compare to the penalty (Halevi, 1999).

7.5 Scheduling

Capacity planning stage *plans* all orders of a company over a long time period and releases a list of orders including delivery date and quantity that *must* be processed at a certain period on the shop floor. Scheduling task is to decide *when* to produce each order. The scheduling problem is formalised as follows:

Given n jobs $J_1, ..., J_n$. of varying times of processing, within every job there is an operation set $O_1, O_2,..., O_n$ which requires scheduling on one of the Mi machines with varying steps of processing, while attempting to meet management objectives which might be:

- Due dates of meetings.
- Minimum levels of work-in-progress nature.
- Maximum processes done.
- Maximum jobs sent from the shop.
- Minimum processes that are completed at a late time.
- Minimum time of waiting of jobs of the shop.
- Minimum jobs number that is waiting in the store.
- Maximum utilisation of job capacity.

More objectives may be added especially from an economic standpoint.

The scheduling problem is an open research topic since 1960. Theoretical methods based on the genetic algorithm, linear programming, branch and bound technique, expert systems, etc., have been proposed. Still, new proposals are being published. Another approach is to construct simple practical dispatching rules such as shortest process time (SPT), first come first serve (FCFS), latest processing time (LPT), Due date based dispatching rule (DDATE), and even random dispatching rule (RANDOM).

It is a common understanding that scheduling is very complex process. IPPS increases tremendously the number of parameters and makes scheduling even more complex. The proposed integration approach is based on the scheduling rules. The mathematical research power within scheduling should be directed towards generating economical rules, such as selecting not the best process, but the one the results with increasing company profit.

The proposed scheduling is based on the following notions:

- Scheduling is maintained by operational resource searches.
- A process plan is variable.
- Priority is given to critical orders.
- Eliminate idle resources.
- Enable alterations in plans of production at any process point.
- Scheduling considers production cost.

7.5.1 Concept and Terminology

Scheduling is in regards to the fact that a resource looks for an operation that is free to perform whenever it is free. A *resource that is free* is clarified as a resource which has completed a function or remains idle and could be instantly loaded. An operation that is free is called as a *technological operation* that could be instantly loaded for process purposes by a computer control algorithm.

The cycle of scheduling commences by scanning every resource when searching for a resource that is free. A resource that is free goes through every operation and the free operation is put down. The best resource operation could be based on the objectives of performance like minimum processing cost or time. The results of scanning with a candidate list for loading using the following rules are:

- Give priority to critical items and all their operations. Use the *look-ahead* function.
- If an item quantity is below xxx attempt to combine orders.
- If the seceding operation of the item that the resource has just finished is (not best) but economical than giving priority.

- If the patch comprises a single entry only, then the functionality is put on the resource.
- If the patch has more than a single entity, then the system puts forth operations with the biggest gap of time for processing it on a resource that is different.
- If there is a list that is vacant, this infers that there is no free best operation which is available for resource processing. Hence, the resource gets to be idle.

*Economical is if time of processing minus a penalty of transfer is lower or equivalent than the best operation time. *Transfer penalty* is called the cost/time of transferring a task to another resource from the previous one. It is inclusive of time of set-up, storage, material handling, etc.

The above are samples of the rule. The user may add more.

7.6 IPPS—Algorithm and Terminology

Scheduling commences with a job list that contains (Halevi, 2004):

1. Item name and number
2. Quantity
3. Process sequence code of priority
4. Critical mark

Process planner prepared a master process for each item as shown in Table 7.4. This master plan has to be adjusted to each of the available resources and in specific order quantity (Halevi, 2014). Two tables; one for minimum cost and one for maximum production are prepared. Such tables are shown in Tables 7.5 and 7.6.

These tables serve as a data supplier for the following algorithm. For programming efficiency, it is recommended to combine all order tables in a single 3D table similar to Table 7.7 – the status of roadmap when R4 is not on duty. For efficiency load the tables with critical mark (means Table 7.5) as first.

The algorithm makes use of the records below:

Resource status file stores the resource status throughout the period of schedule. The data that are stored are:

- Number of resources
- The loaded operation and item

- Resource counter
- Quantity
- The entry sequence number in the file of history.

Resource counter is the counter which puts forth the moment that is remaining for item processing. When put in, it is put forth by multiplying the time of processing by the quantity, as shown in Table 7.7, and it is updated at every cycle of scanning by the moment that has elapsed from the previous cycle of scanning.

History File stores the performance that is actually on the floor of the shop. It keeps the information listed below:

- Sequence number
- Resource number

TABLE 7.7

3D Status of Roadmap When R4 is Idle

Operation	Priority	R1	R2	R3	R4 IDLE	R5	R6	BEST	Δ
Item #3									
001	X	12.5	9.51	5.15	99	4.02	6.54	5	
002	X	5.04	3.93	2.55	99	99	2.82	3	
003	X	6.28	4.86	2.98	2.53	2.47	3.44	5	
004	*0*	*6.38*	*6.12*	*7.05*	*5.78*	*5.93*	*6.83*	*4*	*1.27*
005	4	8.24	6.33	3.67	2.96	2.62	4.42	5	
006	5	5.15	99	4.02	4.86	2.98	2.53	6	
Item #5									
001	X	3.12	3.17	4.02	3.27	99	99	1	
002	*0*	*13.9*	*10.3*	*10.8*	*9.95*	*12.5*	*99*	*4*	*3.95*
003	2	4.86	2.98	2.53	4.86	2.98	2.53	3	
004	2	6.04	4.68	2.90	99	99	3.32	3	
005	4	5.76	4.47	2.8	99	99	3.18	3	
Item #7									
001	X	3.12	3.17	4.02	3.27	99	99	1	
002	X	6.15	4.2	8.05	9.3	99	99	2	
003	*0*	*8.34*	*8.92*	*7.58*	*7.23*	*8.76*	*8.12*	*4*	*1.69*
004	3	2.06	2.11	2.96	2.21	99	99	1	
Item #9									
001	X	4.6	3.60	2.39	99	2.05	2.60	5	
002	X	5.96	4.59	2.87	99	99	3.28	3	
003	*0*	*11.5*	*12.8*	*11.9*	*11.4*	*13.1*	*99*	*4*	*1.7*
004	3	99	99	99	99	1.45	1.72	5	

Note: R1–R6 are the resources numbers.
Bold Italics represents the selected operation number and its values for each item or part.

- Product, operation, and item
- Start time
- Finished time

The file of history's objective is to store information for production and management reports of control. It is incorporated to compare performance to planning that is actual, to get to actual resource load, cost of the item, etc.

The module of scheduling is based on a *sequence cycle* loop which looks into every resource which is put in the *resource status file* then loads resources which are free and upgrades the counter of resources. The sequence cycle loop commences whenever a procedure that was being processed is finished. Here, the resource gets to be idle and the next assignment is when a decision is supposed to be made.

Sequence cycle time is the moment that has passed between the past sequence cycle loop and present time. The time is got from *running clock* which commences at the start of the process of scheduling and advances by the time of working.

7.6.1 Example

The notion behind shop floor control is governed by the availability of a free resource. It means that a free resource will always seek to process an operation which deems to be available. Further, the priority field ("PR") column in 3D (refer Table 7.7) is scanned to identify a free or independent operation. An operation may be categorised as a free operation if PR = 0. Similarly, if the counter-belonging a resource is Zero, the resource is termed as free.

The counter pertaining to field resource along with all other resources are checked and scanned by sequence loop, respectively.

- A Zero counter indicates idleness in last scanned cycle and is considered as it is (check out next scenario)
- In case there exists a non-Zero counter, deduction of time pertaining to the sequence cycle has to be made from resource counter.
- The process pertaining to the present operation is considered to be completed once the result comes out to be Zero. This, further, results in marking the PR of concerned operation as X with the conversion of priorities, belonging to every operation corresponding to this operation number, to 00.
- In an automatic manner, the availability of succeeding operation on the said item is secured and if considered economical, will gain precedence over other operations during processing thereby designating this particular resource for the said operation as the "BEST" or implying the time/cost accounted for processing after deducting the

penalty for transfer is either low or equivalent to the "BEST" time accounted for the said operation. The ensuing step includes allocation of operation to the resource, updated file of status, and setting up of counter to fresh operating timings.

7.6.1.1 Economic Calculations

Table 7.7 shows the status of a shop floor at particular timings. Item #7 of operation 002 was just completed. Operation 003 became free when it was processed on R2. For this operation, the top resource is R4 with 7.23 min per item. A check is carried out if it is economical to process the R2 operation so as to save the time of transfer. 8.92 is the time it takes to process on R2. The time increase is $8.92 - 7.23 = 1.79$ min. Assume that a penalty of transfer of 25 min and quantity is 40 units, then the time increase is $40 \times 1.79 = 71.6$ and the saving is going to be 25 min, then it is said to be non-economical.

Another case: item #9 of operation 002 has just been completed, it was R3 processed, and thus operation 003 was free. The resource that is best for this operation is R4 with 11.4 min per item. A check is carried out if it is economical to process this R3 operation so as to save the time of transfer. The time of processing 11.9 min on R3. The time increase is $11.9 - 11.4 = 0.4$. Assuming that a penalty of transfer of 25 min and a 40-unit quantity, then the time increase is $40 \times 0.4 = 16$ min and the saving is going to be 25 min, then it is economical and the 30th operation is going to be processed by R3.

- In case the succeeding operation allocated to an earlier resource is not found to be economical, or whether the resource remained free in the preceding sequence cycle, a scanning of Table 7.7 in that specific resource is carried out with marking the BEST sign on all the listed operations that are found to be free.

- In case there exists a single entry, the system then allocates the said operation (entry) to the resource. The ensuing step includes updated status file and setting up of counter to fresh operating timings.

- In case there are multiple entries in the list, concentration is paid on to the largest time difference observed while carrying out on a different resource and the operation is allocated by the system accordingly. In order to determine this time difference, the operation row belonging to the pertinent table is scanned and the time gap is computed by calculating the difference between the time of processing taken on the resource that is best and on other different resources. This difference gap is marked on every free operation and the operation with the largest difference value is designated to the resource that is idle in the sequence cycle.

In case there exists an empty list, the "waiting time" pertaining to the best operation to be "free" is searched out by using the *look-ahead* feature. This is carried out by scrutinising the column belonging to the idle resource with regards to an operation which is free. Though such an operation cannot be designated as best, for that particular resource, but as and when the same is stumbled upon, the BEST cell of the corresponding row will point towards the resource that is best. The time of waiting of the resource is indicated by the marking in the resource counter of the field belonging to the *resource status file* (Halevi, 2014).

The above-mentioned procedure has been explained in Table 7.8 depicting the 3D table status at a corresponding stage. As seen, the resource R5 lying idle, thereby a search is made for an operation which is free. Such a

TABLE 7.8

R5 Idle Status

Operation	Priority	Res.-1	Res.-2	Res.-3	Res.-4	Res.-5 IDLE	Res.-6	BEST
Item #3								
001	X	12.5	9.51	5.15	99	4.02	6.54	5
002	X	5.04	3.93	2.55	99	99	2.82	3
003	X	6.28	4.86	2.98	2.53	2.47	3.44	5
004	*0*	*6.38*	*6.12*	*7.05*	*5.78*	*5.93*	*6.83*	*4*
005	4	8.24	6.33	3.67	2.96	2.62	4.42	5
006	5	5.15	99	4.02	4.86	2.98	2.53	6
Item #5								
001	X	3.12	3.17	4.02	3.27	99	99	1
002	*0*	*13.9*	*10.3*	*10.8*	*9.95*	*12.5*	*99*	*4*
003	2	4.86	2.98	2.53	4.86	2.98	2.53	3
004	2	6.04	4.68	2.90	99	99	3.32	3
005	4	5.76	4.47	2.8	99	99	3.18	3
Item #7								
001	X	3.12	3.17	4.02	3.27	99	99	1
002	X	6.15	4.2	8.05	9.3	99	99	2
003	0	8.34	8.92	7.58	7.23	8.76	8.12	4
004	3	2.06	2.11	2.96	2.21	99	99	1
Item #9								
001	X	4.6	3.60	2.39	99	2.05	2.60	5
002	X	5.96	4.59	2.87	99	99	3.28	3
003	0	11.5	12.8	11.9	11.2	13.1	99	4
004	3	99	99	99	99	1.45	1.72	5
	Res.	Item	Op.	Q	Link	Counter	Hist.	
	Res.-4	#2	004	60	22	25	66	
	Res.-1	#7	003	100	23	87	68	

Res., Resource.

Bold Italics represents the selected operation number and its values for each item or part.

free operation (PR = 0). It is observed by further scanning of "BEST" column that the resource R5 is not called by any operation designated as free. The resource marked as R4 is estimated to be BEST when it comes to operation 004 belonging to item #3 which is free as well. Further scanning *resource status file* corresponding to resource item in the row R4 suggests that the ongoing process of operation 004 which requires 25 min more, comes to an end. This indicates that operation 004 has an estimated waiting period of 25 min.

The economic suitability of using an idle resource in order to process an operation designated as free is scrutinised by the system. One procedure involves the calculation of time difference of BEST operation and any alternate operation. This time difference is then compared with the time taken in an idle operation for a free resource. It can be concluded to be economical if the time spent comes out to be smaller to the time gained during this scanning. The procedure is mentioned below:

Processing the operation that is free, operations 004 item #3 is by R4 resource, and it takes 5.78 min per unit. However, the R4 resource will become idle after 25 min. Processing this operation on R5 resource takes 5.93 min on a single unit. Suppose the quantity is 100 units, then by inefficient working and boosting the time of processing by $(5.93 - 5.78) \times 100 = 15$ min gives savings of $(25 - 15) = 10$ min in time of throughput.

The remaining three operations which are open are further checked which concludes it as the best choice. Hence for item #3, resource R5 is subjected to operation 004.

7.6.1.2 Possibility of Disruption

Time of finishing marked on the *file of history* indicates interruption time while the *resource-counter* marked on *resource status file* positioned on 99. It will further be positioned again to Zero in case resource becomes functional. The whole operation (including the item and as well as the operation number) is provided with a new assignment which enlists all the unutilised quantity at its beginning. Be it a single resource disruption or multi, the aforementioned procedure is permissible for both.

7.7 Conclusion

The proposed IPPS system is based on the scheduling rules. The rules might be statesmen's or mathematical algorithms. Computer's power and speed may permit to form any desired rule. This chapter presented sample rules and a Skelton of a computer program which may be prepared in a short time period. The IPPS search algorithms that will reach economic results even by not using the best resource. Using the integration method of dynamic planning, lead time and processing cost are reduced (in the instance of reduction of lead time to 21 from 35 periods and processing cost reduction from 162 to 101 units, refer example presented in Section 7.2.2). The proposed is a practical plan and can eliminate numerous disruptions such as machine break, bottlenecks, new orders, etc., on the shop floor.

7.8 Appendix

However, there is a large field to focus on academic research to validate and create rules. A short list of such research follows:

- When to use maximum production process plan and when minimum cost process plan?
- How to set an hourly rate?
- How to set "penalty" puts forth the extra cost/time to load and unload a resource from an item? It may be an algorithm or a value that is fixed depending on fixtures required, etc.
- Scanning for a resource that is free might result in a candidate list for scheduling. If the list has more than a single entry, then a decision can be made as to which load to be set on the operation.
- Management is supposed to define an algorithm of economic nature to decide on whether to move the function which is next to be processed on the resource, or, which is best or keep it processed on the resource that is original.
- Management is supposed to define the time of idleness; in this case, the resource is supposed to stay idle. Longer management of time is supposed to set rules of economic nature for selecting the function which cancels out time that is idle and becomes economic.
- If there is an empty list, it means that there is no free available function for processing on the resource that is specified. Economic rules to change operation.
- The flexible planning of production treats every order individually. However, in the scheduling stage, a combine jobs rules, in several cases, have an advantage.

Management is supposed to define rules that are appropriate. Instances of rules might be: jobs of varying orders may be combining to a batch of singular nature if they are an item of similar nature of orders that are different and: (i) the time of processing of the batch is less than 60 min, (ii) if the operations that are free at the same time cycle.

References

Halevi, G. *All-Embracing Manufacturing: Roadmap System.* Vol. 59, Heidelberg, NY: Springer Science & Business Media, 2012.

Halevi, G. *Expectations and Disappointments of Industrial Innovations.* Cham: Springer, 2017.

Halevi, G. Global optimization of the manufacturing process. *CIRP International Symposioum – Advanced Design and Manufacturing in the Global Manufacturing Era*, August 21–22, Hong-Kong, 1997, pp. 365–372.

Halevi, G. *Industrial Management-Control and Profit: A Technical Approach.* Lecture Notes in Management and Industrial Engineering. Springer, 2014. ISBN: 978-3-319-03470-6.

Halevi, G. Production management—simple or complex. *International Journal of Innovation and Technology Management*, 1(4), 2004, 359–371.

Halevi, G. *Re-Structuring the Manufacturing Process – Applying the Matrix Method.* The St. Lucie, FL: APICS Series on Resource Management, 1999.

Halevi, G. *Restructuring the Manufacturing Process Applying the Matrix Method.* Vol. 10. CRC Press, 1998. ISBN: 9781574441215.

Halevi, G. *The Effect of Flexibility on Production Scheduling.* Germany: Seminar in Aachen University, 2000.

Halevi, G. *The Role of Computers in Manufacturing Processes.* New York: John Wiley & Sons, Inc., 1980.

Halevi, G. & Kals, H. J. J. Integration of process planning and production management. *International Conference on Computer Applications in Production and Engineering.* Boston, MA: Springer, 1997.

Halevi, G. & Roland, W. *Principles of Process Planning: a Logical Approach.* Dordrecht: Springer Science & Business Media, 2012. doi: 10.1007/978-94-011-1250-5

Halevi, G. & Wang, K. Knowledge based manufacturing system (KBMS). *Journal of Intelligent Manufacturing*, 18(4), 2007, 467–474.

Jain, A., Jain, P. K. & Singh, I. P. An integrated scheme for process planning and scheduling in FMS. *The International Journal of Advanced Manufacturing Technology*, 30(11–12), 2006, 1111–1118.

Jain, A., Jain, P. K., Chan, F. T. & Singh, S. A review on manufacturing flexibility. *International Journal of Production Research*, 51(19), 2013, 5946–5970.

Phanden, R. K., Jain, A. & Verma, R. An approach for integration of process planning and scheduling. *International Journal of Computer Integrated Manufacturing*, 26(4), 2013, 284–302.

Phanden, R. K., Jain, A. & Verma, R. Integration of process planning and scheduling: a state-of-the-art review. *International Journal of Computer Integrated Manufacturing*, 24(6), 2011, 517–534.

8

Due-Date Agreement in Integrated Process Planning and Scheduling Environment Using Common Meta-Heuristics

Halil Ibrahim Demir

Sakarya University

Rakesh Kumar Phanden

Amity University Uttar Pradesh

CONTENTS

8.1 Background and Related Literature 162
    8.1.1 IPPS 163
    8.1.2 SWDDA and Scheduling With Due Window Assignment
          (SWDWA) 163
    8.1.3 IPPSDDA 165
8.2 Problem Definition 166
8.3 Some Common Meta-Heuristics Used in IPPSDDA Problem 169
    8.3.1 Pure Meta-Heuristics 169
          8.3.1.1 RS 169
          8.3.1.2 Simulated Annealing (SA) 170
          8.3.1.3 Semi-Tabu Search (ST) 170
          8.3.1.4 Evolutionary Strategies (ES) 171
          8.3.1.5 GA 171
          8.3.1.6 PSO 172
    8.3.2 Hybrid Meta-Heuristics 174
8.4 An Example From the Literature 174
    8.4.1 Problem Studied 174
    8.4.2 Rules and Formulae Used 175
          8.4.2.1 Due-Date Assignment Rules Used 175
          8.4.2.2 Dispatching Rules Used 176
    8.4.3 Techniques Used 176
    8.4.4 Solutions Compared 176
    8.4.5 Experiments and Results 177
    8.4.6 Discussion of the Results 180
8.5 Summary 180
References 182

8.1 Background and Related Literature

Process planning is a bridging function among design and manufacturing, and it affects the scheduling function in terms of generating an effective and optimal schedule (Lee and Kim, 2001). Developments in Computer Aided Process Planning (CAPP) provide convenience while preparing process plans. This led to the use of alternative process plans in the scheduling. Both process planning and scheduling are important manufacturing functions and they are conventionally treated separately (Zhao et al., 2006). Therefore, schedules are generated using a fixed process plan in a traditional manufacturing system. Since the outputs of process planning become inputs of production scheduling function, these two functions should be integrated to get the global benefit. Additionally, it is better to prepare alternative process plans in order to get a better shop floor workload and higher machine utilisation. Researchers have studied the optimisation of schedule and process plans by IPPS *Integration of Process Planning and Scheduling* to obtain a global benefit. Production scheduling itself belongs to the *Non-Polynomial* hard class of combinatorial optimisation. So, the cohesiveness of both process planning and scheduling functions becomes even harder to solve and it is difficult to find the correct results within a stipulated and realistic duration. Over the last three decades, the IPPS problem has caught researchers' attention and numerous studies have been carried out in literature.

Another important manufacturing function is due-date assignment function. The due dates of products could be calculated (or provided) either internally or externally or through negotiations with customers. If the due-dates are determined externally then firms are responsible to meet these exogenously given due-dates. If the due dates are provided internally or through negotiations with the customers, then the firms will try to determine the best suitable dates for themselves.

Another popular research topic is *Scheduling with Due-Date Assignment* (SWDDA). Here, the scheduling plan and the due-date assignment are solved concurrently but unfortunately, the process planning is not integrated with these two functions. In IPPS research, only process planning, and scheduling functions are integrated but, the due-date assignment function is entirely ignored. In order to increase global benefit, it is better to integrate all three functions, namely, process planning, scheduling, and due-date assignment together. A literature review conducted on IPPS approaches in Chapter 2 suggested that a small number of researches attempted towards IPPSDDA – *Integration of Process Planning, Scheduling and Due-Date Assignment*.

Usually, while solving the IPPS, SWDDA or IPPSDDA problems, only tardiness performance measure is penalised. Besides, as per the JIT – *Just-In-Time* manufacturing philosophy, the earliness performance measure is also undesirable and causes certain costs such as inventory holding, and spoilage cost, etc., to increase.

Demir et al. (2015) criticised earliness, tardiness, and due-date related costs simultaneously in their research. Since due-dates, earliness, and tardiness are all criticised and are undesired, it is better to give the most suitable due-dates and minimise both earliness and tardiness so that the products could be accomplished as per the stipulated due dates. All customers have a degree of importance according to the company and it is better to be more careful with relatively more important customers. Demir and Canpolat (2018) applied weighed due-date assignment and weighted scheduling for the customers. Important customers are assigned relatively close due-dates and scheduled earlier. Thus, substantial savings can be made with this strategy.

8.1.1 IPPS

Process planning is the task of generating a plan for transforming raw material into its finished form according to design specifications (Bhaskaran, 1990). The output of process planning becomes the input to the scheduling function at the shop floor level. On the other hand, scheduling is an important decision-making function which is regularly used by all service and production enterprises. In general, the production scheduling appoints the resources to a task within a given period of time with respect to the optimisation objectives (performance measures) (Pinedo, 2016). Subsequently, scheduling performs apportionments of resources at the factory floor level as per the availability of alternative process plans, it is preferable to provide better input to the downstream scheduling function.

Separately prepared process plans might be equipped without being aware of the dynamic status of the shop floor, or process planner may select the most favoured machines continually, thus it may not pick the other machines. Consequently, the situation leads to disturbed machine loading. Also, this may cause poor shop floor utilisation, and it may not be executable on the floor at all. Classically, both the process planning and scheduling functions work separately and in sequence. Since these two functions are interrelated and one affects the other, in terms of global improvement it is better to integrate these two functions.

Development in computer technology and CAPP provides easier ways to prepare process plans. So, it is easier than before to prepare better and alternative process plans by using technology. Alternative process plans help to obtain better-balanced machine workload at the shop floor level and in case of emergencies or unexpected occurrences they provide a solution to the problem at the shop floor level. Since the IPPS is mentioned throughout the book, it will not be mentioned in detail here.

8.1.2 SWDDA and Scheduling With Due Window Assignment (SWDWA)

Another important function of a manufacturing system is a due-date assignment process. In the last couple of decades, there have been numerous studies on due-date assignment. Afterwards, SWDWA problem also became popular. In the due-date assignment problem, the most suitable due-dates or common due-date are attempted to be determined. In the latter case instead of a single point in time, a window that has a better starting point and a proper length are attempted to be determined. Again, as in due-date assignment, a common due window or separate due windows for each job are attempted to be determined.

Due-dates can be determined externally, internally, or by negotiating with customers. When due-dates are determined externally, firms try to produce according to the predetermined due-dates to optimise performance function. In the second case, the firm determines due-dates and gives due-dates that are most appropriate to the firm's plan and schedule to optimise performance function. In the last case, due-date can be determined by negotiating with the customer. *An important managerial issue in the coordination of the manufacturing-sales interface is the joint determination of order due-dates between customer and manufacturer, mediated by sales personnel* (Lawrence, 1994).

Contrary to the traditional planning which only penalises tardiness, in JIT – *Just in Time* manufacturing philosophy both the earliness, as well as tardiness are penalised. Jobs are desired to be finished neither after nor before the due-dates and they are desired to be finished exactly at their due-dates.

Demir and Erden (2017) penalised due-dates along with the earliness and tardiness. They also used weighted due-date assignment rules in which important customers were given closer due-dates. No one desires long due-dates. A long-due-date means price reduction, loss of customer goodwill, and even worse, it ends up in customer loss. They used weighted due-date assignment, in which important customers are given closer due-dates, which lower the weighted due-date penalty for important customers than the unweighted due-date assignment. In this way, firms keep a good reputation in the eyes of important customers and maintain customer goodwill of valuable customers. Weighted earliness is penalised because of inventory holding, spoilage, insurance, storage of the finished goods and other related costs. Weighted tardiness is penalised because no customer feels happy with tardy jobs, which result in losing the customer goodwill and damages the reputation of a firm in the eyes of customers.

Demir and Canpolat (2018) penalised all weighted due-dates, earliness, and tardiness. The elite customers are provided with convenient due dates and thus schedule their products on priority. Substantial improvements are obtained through the weighted due-date assignment and weighted scheduling in terms of weighed due-date, weighted earliness, and weighted tardiness related costs.

It is better to review the literature survey on SWDDA to obtain more information on the SWDDA problem. Sen and Gupta (1984) and Cheng and Gupta (1989) prepared a survey on SWDDA in the very early times. Later Gordon et al. (2002a) and Lauff and Werner (2004) prepared a survey on SWDDA, where common due-date assignment case was reviewed.

After the SWDDA problem, the SWDWA problem emerged. It is recommended to look at Janiak et al. (2015) to review the SWDWA problem. Some recent researches on SWDWA are as follows: Gerstl and Mosheiov (2013), Xingong and Yong (2015), and Janiak et al. (2015).

In the literature, some studies are on common due-date assignment whereas others are on separate due-date assignments. For example, Gordon et al. (2002a), Gordon and Strusevich (2009), and Tuong et al. (2010) studied scheduling with common due-date assignment problem. On the other hand, Cheng and Kovalyov (1999), (Gordon et al., 2002b), Baykasoğlu et al. (2008), Gordon and Strusevich (2009), Li et al. (2011), and Vinod and Sridharan (2011) studied separate due-date assignments for every job.

There are single machine and multi-machine scheduling studies on SWDDA problem in terms of the manufacturing environment. For a single machine environment, Gordon and Strusevich (2009), Cheng et al. (2011), Li et al. (2011) can be given as examples. For a two-machine flow shop, Birman and Mosheiov (2004) is an example. For a parallel machine environment, Cheng and Kovalyov (1999) can be given as an example. Gupta et al. (2002) and Lauff and Werner (2004) have presented the scheduling of multiple machines and considered the determination of due dates as well. Also, Baykasoğlu et al. (2008) and Vinod and Sridharan (2011) have conducted due date–based job shop scheduling.

8.1.3 IPPSDDA

The process planning, production scheduling, and the due-date assignment are the three key functions of manufacturing that should be integrated to achieve global improvement in a manufacturing system. Each function is defined in the earlier section of this chapter.

Although there are numerous studies on IPPS and SWDDA problems, there are only a few studies on IPPSDDA problem. This study was first performed by Demir and Taskin (2005). They applied genetic algorithms (GA), random search (RS), and hybrid genetic search (RS/GA) methods. Later, Ceven and Demir (2007) investigated the impact of integrating the due-date assignment function into the IPPS problem in an M.S. thesis. These studies are performed in a static shop floor environment.

Demir et al. (2015) studied the integration of SLK (Slack) due-date assignment with IPPS problem and used GAs to observe the contribution of SLK due-date assignment integration instead of externally given due-dates. Demir and Erden (2017) studied the integration of process planning and weighted scheduling with WNOPPT – *Weighted Number of Operation Plus Processing Time* weighted due-date assignment. In this study, they used ordinary solution (OS), RS, evolutionary strategies (ES), GA, hybrid RS/ES, and hybrid RS/GA algorithms. A novel idea is to work on DIPPSDDA – *Dynamic Integrated Process Planning, Scheduling and Due-Date Assignment* problem in order to study the dynamic shop floor environment where jobs are coming dynamically to the shop floor according to an exponential distribution.

If IPPS literature or SWDDA literature are reviewed, it can be seen that weighted due-date assignment is not addressed in them. Demir and his colleagues worked on IPPSDDA problem over a decade and in their researches, they have often used weighted due-date assignment rules. Since weighted due-dates, weighted earliness, and weighted tardiness are all penalised, they suggested to provide closer due-dates to the elite and important customers in order to schedule their products earlier. Implementing this strategy can bring substantial improvements over performance function. Commonly used dispatching rules and due-date assignment rules are presented in Section 8.2.

Demir and Canpolat (2018) studied the integration of WSPT – *Weighted Shortest Processing Time*, scheduling, WSLK – *Weighted Slack* and DDA – *Due-Date Assignment* with process plan selection using GA and ES. Here, again weighted due-date assignment and weighted scheduling approaches are applied. Demir, Canpolat, et al. (2017) applied hybrid semi-tabu search technique to integrate process planning, WSPT – *Weighted Shortest Processing Time*–based scheduling and WSLK – *Weighted slack*–based due-date assignment. Demir, Erden, et al. (2017) studied the process planning problem with WATC – *Weighted Apparent Tardiness Cost*–based scheduling and WPPW – *Weighted Process Plus Weight*–based due-date assignment simultaneously by MDPSO – *Modified Discrete Particle Swarm Optimisation* and SA – *Simulated Annealing* algorithm. Demir, Kokcam, et al. (2017) studied the process planning problem with WEDD – *Weighted Earliest Due Date*–based scheduling and weighted due-date assignment using SA algorithm and ES.

8.2 Problem Definition

An IPPS problem consists of process plan selection among available alternatives and a scheduling problem at the shop floor level. Since only scheduling problem belongs to the NP-Hard class problems and the integrated problem is even harder, early studies on this problem divided the problem into process planning and scheduling subproblems. In the first phase process plan is determined and in the second phase jobs are scheduled (Nasr and Elsayed, 1990; Demir and Wu, 1996).

Later, these phases are combined. Process planning and scheduling are attempted to be solved simultaneously using meta-heuristic algorithms such as GA (Morad and Zalzala, 1999), particle swarm optimisation (PSO) (Guo et al., 2009), and ant colony optimisation (ACO) (Leung et al., 2010; Wang et al., 2014).

SWDDA and SWDWA problems consist of determining due-dates or due-windows and scheduling of the jobs under determined due-dates or due-windows. There are a variety of SWDDA studies and in some of these studies the problem of scheduling with the common due-date assignment, is addressed. In other studies, every job gets its own due date. Some of the studies investigated the SWDDA problem for a single machine environment. In other studies, two machine flow shop scheduling with due-date determination problem is attempted to be solved. There are also studies on parallel machine scheduling with due-date determination. In some of the studies, due-dates are determined in a multi-machine or job shop scheduling environment.

In an IPPSDDA problem, three functions are attempted to be solved concurrently. Better process plans among alternatives are tried to be selected for every job. Every customer is tried to be given better due-date according to the weight of each customer. Thus, important customers are scheduled earlier for closer due-dates and less important customers are scheduled later according to the relatively longer due-dates.

IPPSDDA problem studied by (Demir and Canpolat, 2018) is represented as a chromosome which consists of genes that represent due-date assignment rules, dispatching rules and selected routes of each job. A sample chromosome is represented in Figure 8.1.

Possible due-date assignment rules for weighted and unweighted cases are summarised in Table 8.1, and possible dispatching rules for the weighted and unweighted cases are summarised in Table 8.2.

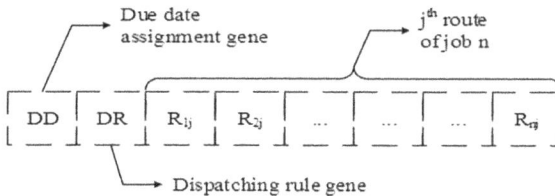

FIGURE 8.1
Chromosome representation of IPPSDDA problem.

TABLE 8.1

Due-Date Assignment Rules

Rule	Explanation	Formula
WCON/CON	(Weighted) constant due-date	$D = (Z_x)d$
WSLK/SLK	(Weighted) slack	$D = p_i + (Z_x)q_x$
WTWK/TWK	(Weighted) total work content	$D = (Z_x)k_x p_i$
WNOP/NOP	(Weighted) number of operations	$D = (Z_x)k_x \text{NOP}$
WNOPPT/NOPPT	(Weighted) number of operations plus processing time	$D = (Z_x)k_x p_i + (Z_y)k_y \text{NOP}$
WPPW/PPW	(Weighted) processing-time-plus-wait	$D = (Z_x)k_x p_i + (Z_y)q_y$
RDM	Random-allowance due-dates	$D = N \sim \left(3 \times P_{\text{avg}}, \left(P_{\text{avg}}\right)^2\right)$

TABLE 8.2

Dispatching Rules

Rule	Explanation	Formula
WATC/ATC	(Weighted) apparent tardiness cost	$\text{Max.} \left\{ I_i(t) = \dfrac{w_i}{p_i} e^{-\frac{\max(d_i - p_i - t, 0)}{KP_{avg}}} \right\}$
WMS/MS	(Weighted) minimum slack	$\text{Min.} \left\{ I_i(t) = \dfrac{\max(d_i - p_i - t, 0)}{w_i} \right\}$
WSPT/SPT	(Weighted) shortest processing time	$\text{Min.} \left\{ I_i(t) = \dfrac{p_i}{w_i} \right\}$
WLPT/LPT	(Weighted) longest processing time	$\text{Max.} \left\{ I_i(t) = \dfrac{p_i}{w_i} \right\}$
WSOT/SOT	(Weighted) shortest operation time	$\text{Min.} \left\{ I_i(t) = \dfrac{p_{ij}}{w_i} \right\}$
WLOT/LOT	(Weighted) longest operation time	$\text{Max.} \left\{ I_i(t) = \dfrac{p_{ij}}{w_i} \right\}$
WEDD/EDD	(Weighted) earliest due-date	$\text{Min.} \left\{ I_i(t) = \dfrac{d_i}{w_i} \right\}$
WERD/ERD	(Weighted) earliest release date	$\text{Min.} \left\{ I_i(t) = \dfrac{r_i}{w_i} \right\}$
SIRO	Service in random order	Jobs are selected randomly

Where;
- Z_x, Z_y change according to the weights of each customer; if the weight is greater than Z_x, Z_y values become appropriately smaller to give closer due-date
- p_i = Total processing time of job i
- d = Unweighted common due-date
- P_{avg} = Mean processing time of all jobs waiting
- k_x = A multiplier such as 1, 2, 3 used in (W)TWK, (W)NOP, (W)NOPPT, and (W)PPW
- q_x = Constant slack added to the processing time in (W)SLK and (W)PPW such as $(P_{avg}/2)$, (P_{avg}) or $(3 \times P_{avg}/2)$

where;
- WATC/ATC = (Weighted) apparent tardiness cost first. This is a composite dispatching rule, and it is a hybrid of MS and SPT
- $I_i(t)$ is priority index of job i
- w_i is the weight of customer i
- p_i is the remaining processing time of job i
- r_i is the release time of job i

- d_i is the due-date of job i
- p_{ij} is the time of j'th operation of job i
- $\max(d_i - p_i - t, 0)$ is i'th job slack
- K is scaling parameter
- P_{avg} is the average processing time of waiting jobs

8.3 Some Common Meta-Heuristics Used in IPPSDDA Problem

As mentioned before, there are only a few studies on IPPSDDA problem and these studies used some common meta-heuristics used in the literature. This section explains the meta-heuristics used in literature for IPPSDDA problem. Pure and hybrid meta-heuristics are used in the solution. Pure meta-heuristics uses only a single specific algorithm. On the other hand, hybrid meta-heuristics utilises two or more meta-heuristics.

8.3.1 Pure Meta-Heuristics

In this section, pure meta-heuristics utilised to solve IPPSDDA in the literature will be investigated. Here, only a single algorithm is used throughout the iterations from the beginning until the end. Demir and Erden (2017) used RS, ES, and GA as a pure solution technique in their research. Demir and Canpolat (2018) again used GA and ES while solving process planning, WSPT scheduling and WSLK due-date assignment concurrently. Demir, Erden, et al. (2017) used modified discrete PSO and simulated annealing algorithms (SA) in their study.

8.3.1.1 RS

Although the RS is not a good search technique for long iterations, it may work well for the initial few iterations. Assume that $X_1, X_2, \ldots, X_n$ are independently and identically distributed uniform random numbers between 0 and 1. Let Y represent the maximum of these n numbers. Expected value of Y can be expressed as: $Y = \max\{X_1, X_2, \ldots, X_n\}$

The cumulative distribution function of Y can be found as follows:

$$F(Y) = P\big(\max(X_1, X_2, \ldots, X_n) \leq y\big) = P\big(X_1 \leq y\big) \times P\big(X_2 \leq y\big), \ldots, P\big(X_n \leq y\big)$$

(Since values of X_i are independent). Since X_i values are uniform between 0 and 1;

$P(X_i \leq y) = y \ (0 \leq y \leq 1)$. F(Y) becomes $F(Y) = y^n$. Then, probability density function becomes $f(y) = n \times y^{(n-1)} (0 \leq y \leq 1)$.

Expected value can be calculated as follows: $E[Y] = \int_0^1 y \times n \times y^{(n-1)} dy = n/(n+1)$

Same can be applied in the minimum case. If we produce random numbers between 0 and 1,000 and take an expected minimum of these numbers, then we get 500, 333, 250, 200 and 167 respectively. Then, the benefits are 500, 157, 83, 50, and 33. Thus, marginal benefit of RS is very high in the very beginning but reduces sharply thereafter. Therefore, it is better to apply an RS at the very first few iterations.

8.3.1.2 Simulated Annealing (SA)

Simulated Annealing algorithm, which is a stochastic search technique, was developed in 1983 by Kirkpatrick et al., (Kirkpatrick et al. 1983). It resembles the annealing process of solids, in which solids are heated to an appropriate temperature for a certain time and are later slowly cooled down. It is better than the descent algorithm where only lower movements are accepted, but worse solutions also have a probability of acceptance to avoid local optimum in the SA algorithm. Increase in performance function by Δ amount can be accepted with probability $P_{accept} = e^{-\left(\frac{\Delta}{T}\right)}$ where (Δ = neighborhood solution − current solution, T = heat). An increase in performance function will be accepted with a higher probability if heat is too high, otherwise P_{accept} will be too low. It is better to start with a higher heat rate and decrease it step by step as the iteration goes on.

8.3.1.3 Semi-Tabu Search (ST)

Tabu search was developed by Glover in 1990, to prevent search end up with a local optimum. Tabu search has a short-term memory to prevent neighbourhood search from being trapped in the local optimum and repeating the same unwanted solutions again and again (Glover 1990). This short-term memory is called a tabu list and this list is updated as iterations go on.

Tabu search is a widely used algorithm in the literature and can be used in IPPS or IPPSDDA problems. On the other hand, instead of tabu algorithm, the semi-tabu algorithm is utilised in the literature to solve IPPSDDA problem. In a shop floor with 100 jobs with 5 alternative routes, there are 102 genes in a chromosome and $10 \times 21 \times 5^{100} = 1.6566079e + 72$ solutions occur. Ina shop floor with 200 jobs with three alternative routes, there are 202 genes in a chromosome and $10 \times 21 \times 3^{200} = 5.5778938e + 97$ solutions occur. So, a tabu list in this vast solution space is very unlikely to occur again. Here ten due-date assignment rules and 21 dispatching rules are assumed to be used.

However, considering the semi-tabu algorithm for IPPSDDA example given in this chapter, there is a possible list of $10 \times 21 = 210$ combinations and it is likely that tabu list should catch unwanted combinations out of 210 possibilities. Thus, the semi-tabu algorithm is quite suitable for the IPPSDDA problem.

Tabu-list contains only dominant genes which consist of due-date assignment rule gene and scheduling rule gene. Remaining genes can be ignored as they are non-dominant and they do not significantly affect the solution. Any changes made in the first two genes dramatically affect the solution performance. Conversely, changes made in the routes of some jobs have little effect on the performance measure as compared to the first two genes.

8.3.1.4 Evolutionary Strategies (ES)

ES was developed by Ingo Rechenberg and Hans-Paul Schwefel, who were students at Berlin Technical University in 1963 (Rechenberg, 1965). Unlike GA, ES uses only mutation operator. While they were working on an optimal flow pattern, they realised that there was a need to conduct many experiments. They continued the experiments by making random changes in parameters that defined the shape, and thus ES was born.

8.3.1.5 GA

GA is a search and optimisation method that is inspired by the evolutionary process observed in nature. In complex multi-dimensional search space, it seeks the best holistic solution based on the principle of survival of the best.

The basic principles of GA were first put forward by John Holland at the University of Michigan. Holland gathered his work in 1975 in his book *Adaptation in Natural and Artificial Systems* (Holland, 1975). First, Goldberg and Holland (1988) used GA for optimisation problems.

GA uses a variety of solutions rather than a single solution in the solution space. Thus, many points are considered at the same time in search space and this increases the probability of a better holistic solution.

GA has been successfully applied in solving problems with a very large search space, where major difficulties are encountered when using other optimisation methods. They do not guarantee to find a holistic best solution to a problem. However, they find acceptable, good solutions within a reasonable amount of time.

As it is stated in the literature, chromosome structure has been used to model the IPPSDDA problem with GA approach. Initially, ten chromosomes that represent the main population are produced. In every iteration, certain pairs of chromosomes are selected for crossover operation and thus crossover population is specified. After that, certain chromosomes are selected for mutation operation and thus mutation population is specified. After applying both crossover and mutation operations, the new main population is acquired from the ex-main population, crossover population and mutation population according to the best performance values. While selecting chromosomes for crossover and mutation operations, performance values of the chromosomes are considered, and better individuals are given higher probability and inferior individuals are given a lower probability.

TABLE 8.3

GA Pseudo-Code

Randomly create main population
BEGIN
n = 0; evaluate population
REPEAT
Create new chromosomes using crossover operator (crossover population)
Create new chromosomes using Mutation operator (mutation population)
Replace main population (best of ex-main, crossover and mutation populations)
n = n + 1
WHILE (n < N)
RETURN $C_{optimum}$ = best of main population as the solution
END

Inferior individuals are possibly selected for crossover and mutation with the hope of getting superior individuals from inferior parents.

Pseudo-code of GA used in IPPSDDA problem is given in Table 8.3.

8.3.1.6 PSO

PSO Algorithm was introduced by Eberhart and Kennedy. It was inspired by the social interaction, communication and social behaviour of bird flocking and fish schooling (Eberhart and Kennedy, 1995). It was first introduced to optimise continuous linear functions. But in the problem studied in IPPSDDA, each gene takes discrete values. These discrete values are integers and categorical values in nature (Strasser et al., 2016). Therefore, PSO is modified to MDPSO algorithm to make it compatible with the IPPSDDA problem (Pan et al., 2008).

In MDPSO algorithm there are particles which constitute a swarm looking for the global optimum. Each particle has performance value, velocity and position, and searches for a better solution. P_{best} represents the personal best solution of each particle throughout the iterations. G_{best} represents the global best solution obtained so far by the swarm. In each iteration, every particle performs three updates. First, some genes of each particle change to another possible value which represents the moment of inertia and at this stage mutation operator is used. After that, certain genes of each chromosome, which are randomly selected, are changed into associated values from P_{best} and each particle moves toward the personal best of that particle so far, and the crossover operator is applied. Finally, every particle moves towards to the G_{best} obtained by the swarm so far and randomly-selected gene values are replaced with the associated values of G_{best} and crossover operation is applied at this stage too.

If S is the swarm size then S particles at iteration k can be expressed as; $X^k = \left[X_1^k X_2^k ... X_S^k \right]$. Then the position of a particle p can be expressed in terms of N genes; $X_p^k = \left[X_{p1}^k X_{p2}^k ... X_{pN}^k \right]$. Personal best of a particle p can be expressed as; $P_{best\,p}^{\ \ k} = \left[P_{p1}^k P_{p2}^k ... P_{pN}^k \right]$. Global Best of the swarm can be expressed as; $G_{best}{}^k = \left[G_1^k G_2^k ... G_N^k \right]$. Velocity of a particle p according to the gene n at iteration k can be expressed as follows;

$$V_{pn}^k = w^{k-1} * V_{pn}^{k-1} + U(0,r1) \otimes \left(P_{best\ pn}^{k-1} - X_{pn}^{k-1} \right) + U(0,r2) \otimes \left(G_{best\ n}^{k-1} - X_{pn}^{k-1} \right)$$

Particle p update can be expressed as follows; $X_p^k = X_p^{k-1} + V_p^k$. Particle p in terms of n genes and its position can be expressed as follows; $X_p = \left[X_{p1} X_{p2} ... X_{pN} \right]$. Particle p can take one of M values for gene n; $X_{pn} = \left[\alpha_{pn}^1 \alpha_{pn}^2 ... \alpha_{pn}^M \right]$. Velocity of particle p in every gene of the chromosome can be expressed as; $V_p = \left[V_{p1} V_{p2} ... V_{pN} \right]$. Particle p's velocity at gene n for state m can be expressed as; $V_{pn} = \left[\delta_{pn}^1 \delta_{pn}^2 ... \delta_{pn}^M \right]$. Demir and his colleagues used equal probability for every possible states of every gene in the chromosome.

In short, modified discrete PSO algorithm used in IPPSDDA problem performs the following three updates; $X_p^k = U(0,r2) \otimes f3 \big(U(0,r1) \otimes$ $f2 \big(U(0,w) \otimes f1 \big(X_p^{k-1} \big), P_{best\ p}^{k-1} \big), G_{best}{}^{k-1} \big)$

First, $f1$ function is applied and with probability w $(0 \leq w \leq 1)$ selected genes of the current chromosome are changed to another possible state by applying mutation operator. After that, $f2$ function is applied and new current chromosome and P_{best} chromosome are crossed over and with probability $r1$ $(0 \leq r1 \leq 1)$ new chromosome assumes the value of P_{best} chromosome for the selected genes. Finally, $f3$ function is applied. The newly updated chromosome is crossed over with G_{best} chromosome with probability $r2$ $(0 \leq r2 \leq 1)$. Finally, new chromosome assumes the value of G_{best} for the selected genes. The pseudo-code of the MDPSO algorithm used in IPPSDDA problem is given in Table 8.4.

TABLE 8.4

MDPSO Pseudo-Code

Randomly create $P_{current}$ population
$P_{best} = P_{current}$
BEGIN
n = 0; evaluate population
G_{best} (chromosome) = best of P_{best} population
REPEAT
n = n + 1;
  FOR (every chromosome (j) in $P_{current}$ population) DO
    Apply mutation operator to chromosome (j)
    Change randomly selected genes with associated P_{best} values
    Change randomly selected genes with associated G_{best} values
    Evaluate new chromosomes
    If new chromosome (j) < P_{best} (j)
      P_{best} (j) = new chromosome (j)
    If best of P_{best} population < G_{best}
      G_{best} (chromosome) = best of P_{best} population
WHILE (n < N)
RETURN G_{best} as the solution
END

8.3.2 Hybrid Meta-Heuristics

In this section, hybrid meta-heuristics used in the literature while solving the IPPSDDA problem are explained. Hybrid meta-heuristics employ two or more algorithms concurrently throughout the iterations from the beginning until the end. (Demir and Canpolat, 2018) used Hybrid RS/ES, and Hybrid RS/GA algorithms while solving their problems. Demir, Canpolat, et al. (2017) used Hybrid (RS/ST) in their solution. Demir, Kökçam, et al. (2017) used RS/SA hybrid and RS/ES hybrid while solving their IPPSDDA problem.

8.4 An Example From the Literature

In this section, an example from the literature on IPPSDDA is presented, which is a study by Demir and Erden (2017) titled "Solving process planning and weighted scheduling with WNOPPT weighted due-date assignment problem using some pure and hybrid meta-heuristics". In this study, weighted scheduling and weighted due-date assignment rules are used. Scheduling and due-date assignment rules used in this study are further explained in this section.

8.4.1 Problem Studied

In this example, weighted scheduling, WNOPPT weighted due-date assignment functions are integrated with process plan selection. Eight different shop floors are studied as described in Table 8.5.

TABLE 8.5

Shop Floor Configurations

Shop Floor No.	#of Jobs	# of Machines	# of Routes	# of Op. per Job	Processing Times
1	25	5	5	10	$(10 + z * 6)$
2	50	10	5	10	$(10 + z * 6)$
3	75	15	5	10	$(10 + z * 6)$
4	100	20	5	10	$(10 + z * 6)$
5	125	25	3	10	$(10 + z * 6)$
6	150	30	3	10	$(10 + z * 6)$
7	175	35	3	10	$(10 + z * 6)$
8	200	40	3	10	$(10 + z * 6)$

To understand this table, let us look at the shop floor 2, which has 50 jobs, ten machines, each job has five alternative routes, and every operation has normally distributed time according to the $(10 + z \times 6)$ formula.

8.4.2 Rules and Formulae Used

In this study, weighted due-date, earliness and tardiness are all penalised. It is assumed that the shop floor works a single shift per day, and it takes $8 \times 60 = 480$ min.

$$PD(j) = weight\,(j) \times 8 \times \left(\frac{D}{480}\right)$$

$PD(j)$ is the penalty for the due-date of job j (1)

$$PE(j) = weight(j) \times \left(5 + 4 \times \left(\frac{E}{480}\right)\right)$$

$PE(j)$ is the penalty for the earliness of job j (2)

$$PT(j) = weight\,(j) \times \left(10 + 12 \times \left(\frac{T}{480}\right)\right)$$

$PT(j)$ is the penalty for the tardiness of job j (3)

$$Penalty\,(j) = PD\,(j) + PE\,(j) + PT\,(j)$$

Penalty (j) is the total penalty of job j that includes due-date, earliness and tardiness related costs (4)

$$Total\ Penalty = \sum_{j} Penalty\,(j)$$

Total penalty incurred for all the jobs (5)

8.4.2.1 Due-Date Assignment Rules Used

In this study, mainly two due-date assignment rules are utilised. Using the multipliers, they have become ten rules as given in Table 8.6. First nine rules are weighted WNOPPT due-date assignment rules. The last rule represents random due-date assignment and can be considered as external due-dates.

TABLE 8.6

Due-Date Assignment Rules

Method	Multiplier 1	Multiplier 2	Rule No.
WNOPPT	$k_x = 1, 2, 3$	$k_y = 1, 2, 3$	1, 2, 3, 4, 5, 6, 7, 8, 9
RDM			10

TABLE 8.7

Dispatching Rules

Method	Multiplier	Rules
WATC, ATC	$k_x = 1, 2, 3$	1, 2, 3, 4, 5, 6
WMS, MS		7, 8
WSPT, SPT		9, 10
WLPT, LPT		11, 12
WSOT, SOT		13, 14
WLOT, LOT		15, 16
WEDD, EDD		17, 18
WERD, ERD		19, 20
SIRO		21

8.4.2.2 Dispatching Rules Used

Mainly nine different dispatching rules are utilised. There are totally 21 rules generated considering weights and multipliers, and all rules are summarised in Table 8.7.

8.4.3 Techniques Used

Five search techniques as mentioned earlier and OS are compared. As pure meta-heuristics, RS, ES and GA are used. As hybrid searches, hybrid ES and hybrid GA are applied.

Hybrid ES and hybrid GA start with an undirected search and later iterations become directed. In order to scan solution space better and faster, initially, the RS is applied in these hybrids.

The OS is the first randomly created chromosome before any further iterations are applied. Obtained results are compared to see how well each search algorithm performs.

8.4.4 Solutions Compared

In this study, four levels of integration are compared. In each level, six solutions are compared. At the first level, all functions are disintegrated. Due-dates are determined randomly, and jobs are scheduled in random order. At this level, SIRO-RDM (OS, GA, RS/GA, ES, RS/ES, RS) solutions are found and compared within the level and with higher integration levels. After that WSCH-RDM (OS, GA, RS/GA, ES, RS/ES, RS) solutions are found and compared with other levels. In this level, due-dates are still determined randomly but jobs are scheduled according to a good dispatching rule among 21 scheduling rules and many of them consider the weights of each customer. Later SIRO-WNOPPT (OS, GA, RS/GA, ES, RS/ES, RS) solutions are found and compared. In this level, due-dates are determined according to

TABLE 8.8

Iteration Numbers for Pure and Hybrid Searches

	ES	RS/ES Hybrid		RS	GA	RS/GA Hybrid	
Shop Floor	**ES Iter#**	**Random Iter#**	**ES Iter#**	**Random Iter#**	**GA Iter#**	**Random Iter#**	**GA Iter #**
1, 2	200	20	180	200	200	20	180
3, 4	150	15	135	150	150	15	135
5, 6	100	10	90	100	100	10	90
7, 8	50	5	45	50	50	5	45

Iter, Interation.

WNOPPT that considers each customer's weight while assigning due-dates, but jobs are scheduled randomly. Lastly, full integration level which is the most important level is solved, and WSCH-WNOPPT (OS, GA, RS/GA, ES, RS/ES, RS) solutions are found for comparison. At this level, weighted due-date assignment and weighted scheduling are integrated with process plan selection. Iteration sizes for each algorithm are summarised in Table 8.8. OS, RS, GA, ES and two hybrid searches are applied at each level.

8.4.5 Experiments and Results

Eight shop floors were sequentially tested with four integration levels and the results of selected four shop floors are summarised in Table 8.9. A comparison of the methods in the highest level of integration is given in Figures 8.2–8.5. Firstly, the totally unintegrated problem is tested and later every function is integrated step by step and the results considering 24 solutions for each level are summarised below;

At the first level of integration where randomness plays a very important role, RS/ES gave best solutions two times, ES gave best solutions two times and GA gave best solutions four times. At the second level where scheduling is integrated with process plan selection: ES gave the best solutions four times, RS/ES gave best solutions twice and RS/GA gave the best solutions twice. This level was always better than the first level and substantial improvements are obtained in every eight shop floors. At the third level, the process plan selection has been integrated with the due-date assignment, while the parts are dispatched through a casual direction. In this level, RS/GA gave the best solutions five times and GA gave the best solutions three times. For every eight shop floors, this level is always found to be better than the first level. Although the SIRO scheduling severely deteriorates the performance at this level, there are still substantial improvements obtained as compared to the unintegrated first level.

At the fourth level, all the three functions are integrated and this is the highest integration level that always found the best solutions for all levels for all eight shop floors. The most drastic improvements are obtained at this

TABLE 8.9

Comparative Evaluation of 24 Combinations on Four Shop Floors

Level of Integration	Shop Floor											
	2			4			6			8		
	B	A	W	B	A	W	B	A	W	B	A	W
1-SIRO-RDM-OS	611	611	611	1,337	1,337	1,337	1,724	1,724	1,724	2,490	2,490	2,490
1-SIRO-RDM-ES	552	564	568	1,201	1,208	1,213	1,596	1,608	1,615	2,309	2,344	2,351
1-SIRO-RDM-RS/ES	523	533	539	1,201	1,224	1,231	1,694	1,715	1,737	2,307	2,325	2,331
1-SIRO-RDM-GA	535	540	543	1,201	1,219	1,224	1,579	1,584	1,587	2,273	2,288	2,295
1-SIRO-RDM-RS/GA	545	549	553	1,178	1,183	1,188	1,603	1,611	1,619	2,277	2,293	2,303
1-SIRO-RDM-RS	558	565	571	1,254	1,261	1,266	1,610	1,645	1,657	2,346	2,367	2,378
2-WSCH-RDM-OS	560	560	560	1,214	1,214	1,214	1,621	1,621	1,621	2,280	2,280	2,280
2-WSCH-RDM-ES	416	420	422	1,009	1,018	1,024	1,263	1,271	1,275	1,808	1,828	1,833
2-WSCH-RDM-RS/ES	430	438	440	965	971	976	1,330	1,513	1,652	1,835	1,847	1,851
2-WSCH-RDM-GA	441	446	450	989	998	1,004	1,286	1,287	1,287	1,828	1,831	1,834
2-WSCH-RDM-RS/GA	423	424	425	957	959	961	1,267	1,269	1,270	1,824	1,825	1,826
2-WSCH-RDM-RS	458	462	464	997	1,014	1,030	1,319	1,338	1,357	1,906	1,943	1,968
3-SIRO-WNOPPT-OS	609	609	609	1,243	1,243	1,243	1,627	1,627	1,627	2,283	2,283	2,283
3-SIRO-WNOPPT-ES	513	524	531	1,138	1,179	1,190	1,530	1,540	1,549	2,145	2,163	2,176
3-SIRO-WNOPPT-RS/ES	527	530	533	1,170	1,185	1,195	1,579	1,639	1,713	2,162	2,176	2,187
3-SIRO-WNOPPT-GA	487	495	501	1,123	1,136	1,141	1,507	1,512	1,516	2,141	2,152	2,161
3-SIRO-WNOPPT-RS/GA	491	497	499	1,115	1,128	1,134	1,503	1,522	1,529	2,087	2,116	2,134
3-SIRO-WNOPPT-RS	511	522	528	1,177	1,187	1,195	1,528	1,553	1,562	2,173	2,210	2,223
4-WSCH-WNOPPT-OS	488	488	488	962	962	962	1,265	1,265	1,265	1,774	1,774	1,774
4-WSCH-WNOPPT-ES	360	364	367	846	854	859	1,093	1,104	1,111	1,602	1,611	1,617
4-WSCH-WNOPPT-RS/ES	357	361	364	852	857	862	1,176	1,288	1,570	1,633	1,650	1,657
4-WSCH-WNOPPT-GA	402	405	406	845	851	853	1,065	1,069	1,073	1,623	1,629	1,632
4-WSCH-WNOPPT-RS/GA	398	399	400	847	854	857	1,065	1,074	1,077	1,565	1,571	1,575
4-WSCH-WNOPPT-RS	414	420	423	892	901	910	1,119	1,134	1,142	1,626	1,666	1,689

B, Best; A, Average; W, Worst.

FIGURE 8.2
Shop floor 2 (50 × 10 × 5).

FIGURE 8.3
Shop floor 4 (100 × 20 × 5).

FIGURE 8.4
Shop floor 6 (150 × 30 × 3).

FIGURE 8.5
Shop floor 8 (200 × 40 × 3).

level and results prove that the highest integration is the best level of integration. GA gave the best results in five out of eight shop floors at this level. Whereas at shop floors two and three, RS/ES gave the best results, and at the shop floor eight, RS/GA gave the best results. Thus, GA is found as the best method, but hybrid searches also found other promising methods. In hybrid searches, 10% of the iterations are consumed randomly. RS is useful at the very beginning but only for the first few iterations. After that, it becomes a marginally inefficient technique.

8.4.6 Discussion of the Results

In this study, the IPPSDDA problem is comprehensively discussed. Weighted scheduling and weighted WNOPPT due-date assignment rules are applied. Four levels of integration are tested. Initially, the unintegrated level is solved and this was the poorest level as expected. After that, firstly weighted scheduling, and later weighted due-date assignment is integrated with process plan selection. These intermediate integration levels are found to be substantially useful, especially the weighted scheduling integration. The weighted due-date assignment is also substantially useful but SIRO rules deteriorate the performance severely, therefore this level of integration appears poorer than it really is. Finally, all the three functions are integrated, and this highest level of integration is found to be the most successful level in terms of performance function as expected. This level proves how IPPSDDA integration is useful.

8.5 Summary

Traditionally, process planning, scheduling and due-date assignment functions were not integrated altogether and were treated separately. Although there are numerous researches on IPPS and SWDDA, only a few researches are on IPPSDDA. Due-dates can be determined internally, externally, or by negotiating with the customers and determined with mutual agreement. IPPS researches proved how integration is useful, but unfortunately, in these researches, due-dates are not determined concurrently. In the other integration problem, SWDDA, process planning is not integrated with scheduling and due-date assignment. The purpose of this study is to obtain higher global performance by integrating all the three functions as compared to intermediate integration levels.

Developments in technology, hardware, software and algorithms have made it possible to solve previously unsolvable problems, and have eased the process of finding solutions to difficult problems. Developments in technology help us to prepare process plans more easily as compared to previous instances. Traditionally, only tardiness is penalised while scheduling jobs. But, according to JIT manufacturing philosophy, both earliness and tardiness should be penalised. Demir and his colleagues penalised due-dates as well, to abstain long due-dates. Due-dates can be determined exogenously. In this case, firms try to obtain better performance according to these external due-dates. In other cases, firms may determine due-dates internally or after negotiating with customers. In these cases, firms try to determine better due-dates that help to obtain better performance.

According to literature, SWDDA problem has been studied in different environments. Some of the studies tackled the problem using single machine environment, some of them using two-machine flow shop environment, some of them using parallel machine environment, and some others using the multi-machine environment, and job shop scheduling environment.

Only scheduling problem falls into NP-Hard problem class, and integrated problems are even harder to solve. Thus, many meta-heuristics have been applied to the solution of integrated problems. Nowadays numerous meta-heuristics are developed and applied to solve IPPS, SWDDA and IPPSDDA problems. Some of these meta-heuristics applied in the solution of IPPSDDA problem are summarised in this chapter of the book.

In the literature sometimes common due-dates are assigned for the jobs in the batch delivery case or for the parts waiting to be assembled. CON due-date assignment rule is applied in common due-date assignment case. In some other researches, each job gets its own due-date and some prevalent due-date assignment rules are applied such as SLK, TWK, PPW, NOP, etc.

Demir and his colleagues recently used weighted due-date assignment rules such as WSLK, WTWK, WPPW, WNOPPT, etc. In these studies, closer due-dates are given to relatively more important customers, and longer due-dates are given to less important customers. Using weighted due-date assignment, substantial improvements are aimed in terms of penalties in weighted due-date, weighted earliness, and weighted tardiness related costs.

Finally, Demir and his colleagues tested every level of IPPSDDA. First of all, the unintegrated level is tested, and jobs are scheduled in random order and due-dates are determined randomly. As expected, this level is found to be the worst level. Later, intermediate integration levels are tested. First, process planning is integrated with scheduling. Second, process planning is integrated with due-date assignment. In these levels, substantial improvements are obtained. Finally, all three functions are integrated by using common meta-heuristics such as GA, SA, ST, ES, MDPSO, RS and using some hybrid algorithms such as RS/GA, RS/ES, RS/SA, RS/ST, RS/MDPSO, GA/MDPSO and RS/GA/MDPSO algorithms. In these hybrid algorithms, RS is found useful at the very early few iterations. After a few iterations, RS appeared to be time consuming and insufficient compared to the other common meta-heuristics.

Full integration level, which integrates process planning, scheduling and due-date assignment, is found to be the most successful level and the highest improvements are achieved in this fully integrated level, as expected.

All of the IPPSDDA researches are studied in a job shop environment, especially for the static shop floor case. As future studies, IPPSDDA problem can be studied in different machining environments as on dynamic shop floor case. Also, other recent meta-heuristics can be applied such as ACO and ABC – *Artificial Bee Colony* algorithm, etc.

References

Baykasoğlu, A., Göçken, M., Unutmaz, Z.D., 2008. New approaches to due date assignment in job shops. *European Journal of Operational Research* 187, 31–45.

Bhaskaran, K., 1990. Process plan selection. *The International Journal of Production Research* 28, 1527–1539.

Birman, M., Mosheiov, G., 2004. A note on a due-date assignment on a two-machine flow-shop. *Computers and Operations Research* 31, 473–480.

Ceven, E., Demir, H.I., 2007. Benefits of integrating due-date assignment with process planning and scheduling. Master of Science Thesis, Sakarya University.

Cheng, T.C.E., Gupta, M.C., 1989. Survey of scheduling research involving due date determination decisions. *European Journal of Operational Research* 38, 156–166.

Cheng, T.C.E., Kovalyov, M.Y., 1999. Complexity of parallel machine scheduling with processing-plus-wait due dates to minimize maximum absolute lateness. *European Journal of Operational Research* 114, 403–410.

Cheng, T.E., Cheng, S.-R., Wu, W.-H., Hsu, P.-H., Wu, C.-C., 2011. A two-agent single-machine scheduling problem with truncated sum-of-processing-times-based learning considerations. *Computers & Industrial Engineering* 60, 534–541.

Demir, H.I., Canpolat, O., 2018. Integrated process planning, WSPT scheduling and WSLK due-date assignment using genetic algorithms and evolutionary strategies. *An International Journal of Optimization and Control: Theories & Applications* 8(1), 73–83.

Demir, H.I., Canpolat, O., Uygun, O., Goksu, A., 2017. Integrating process planning, weighted shortest processing time scheduling and weighted slack due-date assignment using hybrid semi-tabu search, *19th International Conference on Emerging Trends in Engineering Technology.* pp. 285–285.

Demir, H.I., Erden, C., 2017. Solving process planning and weighted scheduling with WNOPPT weighted due-date assignment problem using some pure and hybrid meta-heuristics. *Sakarya University Journal of Science* 21, 210–222.

Demir, H.I., Erden, C., Uygun, Ö., Kokcam, A.H., 2017. Solving process planning WATC scheduling and WPPW due date assignment concurrently using modified discrete PSO and simulated annealing algorithms, *Proceedings 4th International Conference on Computational and Experimental Science and Engineering (ICCESEN2017)*. p. 472.

Demir, H.I., Kokcam, A.H., Simsir, F., Uygun, O., 2017. Solving process planning, weighted earliest due date scheduling and weighted due date assignment using simulated annealing and evolutionary strategies. World academy of science, engineering and technology. *International Journal of Industrial and Manufacturing Engineering* 11, 1512–1519.

Demir, H.I., Taskin, H., 2005. Integrated process planning, scheduling and due-date assignment. Ph.D. Thesis, Sakarya University.

Demir, H.I., Uygun, O., Cil, I., Ipek, M., Sari, M., 2015. Process planning and scheduling with SLK due-date assignment where earliness, tardiness and due-dates are punished. *Journal of Industrial and Intelligent Information* 3(3), 173–180.

Demir, H.I., Wu, S.D., 1996. A comparison of several optimization schemes for the integrated process planning and production scheduling problems. Master of Science Thesis, Lehigh University.

Eberhart, R., Kennedy, J., 1995. A new optimizer using particle swarm theory, *MicroMachine Human Science, 1995. MHS'95., Proceedings Sixth International Symposium*, Nagoya, Japan. IEEE, pp. 39–43.

Gerstl, E., Mosheiov, G., 2013. An improved algorithm for due-window assignment on parallel identical machines with unit-time jobs. *Information Processing Letters* 113, 754–759.

Glover, F., 1990. Tabu search: A tutorial. *Interfaces* 20, 74–94.

Goldberg, D.E., Holland, J.H., 1988. Genetic algorithms and machine learning. *Machine Learning* 3, 95–99.

Gordon, V., Proth, J.M., Chu, C., 2002a. A survey of the state-of-the-art of common due date assignment and scheduling research. *European Journal of Operational Research* 139, 1–25.

Gordon, V., Proth, J.M., Chu, C., 2002b. Due date assignment and scheduling: SLK, TWK and other due date assignment models. *Production Planning & Control* 13, 117–132.

Gordon, V.S., Strusevich, V.A., 2009. Single machine scheduling and due date assignment with positionally dependent processing times. *European Journal of Operational Research* 198, 57–62.

Guo, Y.W., Li, W.D., Mileham, A.R., Owen, G.W., 2009. Optimisation of integrated process planning and scheduling using a particle swarm optimisation approach. *International Journal of Production Research* 47, 3775–3796.

Gupta, J.N.D., Krüger, K., Lauff, V., Werner, F., Sotskov, Y.N., 2002. Heuristics for hybrid flow shops with controllable processing times and assignable due dates. *Computers and Operations Research* 29, 1417–1439.

Holland, J.H., 1975. Adaptation *in Natural and Artificial Systems: An Introductory Analysis with Applications to Biology, Control, and Artificial Intelligence*. Ann Arbor, MI: University of Michigan Press.

Janiak, A., Janiak, W.A., Krysiak, T., Kwiatkowski, T., 2015. A survey on scheduling problems with due windows. *European Journal of Operational Research* 242, 347–357.

Kirkpatrick, S., Gelatt, C.D., Vecchi, M.P., 1983. Optimization by simulated annealing. *Science* 220, 671–680.

Lauff, V., Werner, F., 2004. Scheduling with common due date, earliness and tardiness penalties for multimachine problems: A survey. *Mathematical and Computer Modelling* 40, 637–655.

Lawrence, S.R., 1994. Negotiating due-dates between customers and producers. *International Journal of Production Economics* 37, 127–138.

Lee, H., Kim, S.S., 2001. Integration of process planning and scheduling using simulation based genetic algorithms. *The International Journal of Advanced Manufacturing Technology* 18, 586–590.

Leung, C.W., Wong, T.N., Mak, K.-L.L., Fung, R.Y.K., 2010. Integrated process planning and scheduling by an agent-based ant colony optimization. *Computers and Industrial Engineering* 59, 166–180.

Li, S., Ng, C.T., Yuan, J., 2011. Group scheduling and due date assignment on a single machine. *International Journal of Production Economics* 130, 230–235.

Morad, N., Zalzala, A.M.S., 1999. Genetic algorithms in integrated process planning and scheduling. *Journal of Intelligent Manufacturing* 10, 169–179.

Nasr, N.A.B.I.L., Elsayed, E.A., 1990. Job shop scheduling with alternative machines. *International Journal of Production Research* 28, 1595–1609.

Pan, Q.-K., Tasgetiren, M.F., Liang, Y.-C., 2008. A discrete particle swarm optimization algorithm for the no-wait flowshop scheduling problem. *Computers & Operations Research* 35, 2807–2839.

Pinedo, M.L., 2016. *Scheduling: Theory, Algorithms, and Systems*. New York, Dordrecht, Heidelberg and London: Springer.

Rechenberg, I., 1965. *Cybernetic Solution Path of an Experimental Problem*, Vol. 1122. Britain: Ministry of Aviation, Royal Aircraft Establishment, Library Translation.

Sen, T., Gupta, S.K., 1984. A state-of-art survey of static scheduling research involving due dates. *Omega* 12, 63–76.

Strasser, S., Goodman, R., Sheppard, J., Butcher, S., 2016. A new discrete particle swarm optimization algorithm, *Proceedings 2016 Genetic Evolutionary Computation Conference – GECCO textquotesingle16*, Denver, CO. ACM Press.

Tuong, H.N., Soukhal, A., Billaut, J.C., 2010. A new dynamic programming formulation for scheduling independent tasks with common due date on parallel machines. *European Journal of Operational Research* 202, 646–653.

Vinod, V., Sridharan, R., 2011. Simulation modeling and analysis of due-date assignment methods and scheduling decision rules in a dynamic job shop production system. *International Journal of Production Economics* 129, 127–146.

Wang, J., Fan, X. F., Zhang, C., Wan, S., 2014. A graph-based ant colony optimization approach for integrated process planning and scheduling. *Chinese Journal of Chemical Engineering* 22(7), 748–753.

Xingong, Z., Yong, W., 2015. Single-machine scheduling CON/SLK due window assignment problems with sum-of-processed times based learning effect. *Applied Mathematics and Computation* 250, 628–635.

Zhao, F., Zhu, A., Ren, Z., Yang, Y., 2006. Integration of process planning and production scheduling based on a hybrid PSO and SA algorithm, *Proceedings 2006 IEEE International Conference on Mechatronics and Automation*, Denver, CO. IEEE, pp. 2290–2295.

9

Integration of Process Planning and Scheduling: An Approach Based on Ant Lion Optimisation Algorithm

Milica Petrović and Zoran Miljković

University of Belgrade

CONTENTS

9.1   Introduction 185
9.2   Flexibility Types and Representation of IPPS Solutions 187
9.3   The Mathematical Modelling of Objective Functions 188
9.4   ALO Algorithm 190
9.5   Implementation Procedure of ALO Algorithm for IPPS
        Optimisation 194
9.6   Experimental Verification 194
        9.6.1   Experiment 1 195
        9.6.2   Experiment 2 198
9.7   Conclusion 201
Acknowledgement 203
References 203

9.1 Introduction

Computer-aided process planning (CAPP) system was introduced during the last decade of the 20th century in order to bridge the gap between two manufacturing functions, i.e., computer-aided design (CAD) and computer-aided manufacturing (CAM). This system is well-known to develop systematic procedures to generate an optimal process plan for part manufacturing, by taking into consideration available manufacturing resources as well as constraints related to costs and precedence relations. On the other hand, the scheduling function can be described as time allocation of manufacturing operations, with regards to alternative manufacturing resources. After receiving process plans of the parts as inputs, scheduling function outputs a sequence of operations performed by using selected manufacturing resources (machine tools and cutting tools). In traditional approaches to model process planning and scheduling, these two functions have been performed separately and sequentially. However, various requirements such as increase in the utilisation level of manufacturing utilities, reduction in production costs, and elimination in production bottlenecks motivates the research towards the development of an effective approach to simultaneously deal with the two manufacturing functions of process planning and scheduling. Since these manufacturing functions are complementary and interrelated, many researchers have developed and proposed different methods for "integration of process planning and scheduling" (IPPS) to increase the performance of the entire production system.

Guo et al. (2009) formulated IPPS problem as follows: *Given a set of N parts which are to be processed on machines with operations including alternative manufacturing resources, select suitable manufacturing resources and sequence the operations so as to determine a schedule in which the precedence constraints among operations can be satisfied and the corresponding objectives can be achieved.*

In the manufacturing system, integration of these two functions with available alternative resources classifies IPPS into the group of NP-hard (non-deterministic polynomial) combinatorial problem of optimisation. During the last decade, swarm intelligence and evolutionary algorithms have been considered a good alternative to classical optimisation methods when optimisation problems are multi-dimensional, nonlinear, and complex. Biologically inspired methods such as meta-heuristic algorithms and artificial intelligence techniques find their successful applications in various engineering problems and fields such as process planning (Petrović et al. 2015, 2016a; Miljković & Petrović 2017), learning of robot motion trajectories (Vuković et al. 2015; Mitić & Miljković 2015; Mitić et al. 2018), visual control of robotic systems (Miljković et al. 2013a,b; Mitić & Miljković 2014), prediction of robot failures (Diryag et al. 2014), autonomous navigation (Babić et al. 2012), optimisation (Mitić et al. 2015), and neural networks applications (Vuković et al. 2018).

The state-of-the-art for IPPS includes methodologies based on the following approaches: genetic algorithm (GA) (Morad & Zalzala 1999; Lee & Kim 2001; Shao et al. 2009; Petrović et al. 2012; Phanden et al. 2013; Dai et al. 2015; Zhang et al 2016; Lee & Ha 2019); simulated annealing (Li & McMahon 2007); symbiotic evolutionary algorithm (Kim et al. 2003); particle swarm optimisation (PSO) (Guo et al. 2009); chaotic particle swarm optimisation (Petrović et al. 2016c), ant colony optimisation (ACO) (Zhang & Wong 2018); agent-based approach (Wong et al. 2006a,b; Li et al. 2010); algorithm inspired by imperialist competition (Lian et al. 2012); improved GA (Zhang & Yan 2005; Lihong & Shengping 2012); approach based on game theory and hybrid GA-SA algorithm (Li et al. 2012); neuro-fuzzy model (Seker et al. 2013); decomposition algorithm (Barzanji et al. 2019); approach based on tabu search and grammars (Baykasoğlu & Özbakır 2009); hybrid multi-agent GA (Petrović et al. 2013); hybrid agent-based two-stage ACO approach (Wong et al. 2012); hybrid GA and PSO (Yu et al. 2015; Li et al. 2019); hybrid ACO and PSO algorithms (Srinivas et al. 2012).

The research in this chapter is aimed at investigating the implementation of nature-inspired Ant Lion Optimisation (ALO) technique for optimal IPPS functions. The main motivation for the application of ALO algorithm comes from experimental results presented by Mirjalili (2015), where this algorithm showed advantages in improving exploration and exploitation search abilities, convergence rate characteristics, as well as the capability to avoid the local optima.

The succeeding sections of the present chapter are structured as follows. Section 9.2 introduces IPPS and types of manufacturing flexibility as well as the representation methods of IPPS problem. Section 9.3 presents the three objective functions along with mathematical modelling utilised for the optimisation process. The proposed ALO algorithm is introduced in Section 9.4, while the implementation methodology for the IPPS functions is presented in Section 9.5. In order to validate the proposed ALO methodology, comprehensive experimental evaluation by using 25 benchmarks is performed in Section 9.6, which also contains the comparative results of GA and PSO algorithm achieved through two studies. Section 9.7 is devoted to a summary of concluding remarks.

9.2 Flexibility Types and Representation of IPPS Solutions

The present research work emphases on the four kinds of manufacturing flexibilities involved in IPPS problem visualisation – processing flexibility, tool flexibility, machine flexibility, and sequencing flexibility.

- In processing flexibility, the aim is to explore the possibility of performing a particular operation by the available alternative sequences of operations or an operation to manufacture the same part.
- Tool flexibility relates to using alternative cutting tools for the same manufacturing operation.
- Machine flexibility provides the possibility of using alternative machine tools to perform the same manufacturing operation.
- Sequencing flexibility assumes that manufacturing operations can be interchanged in order to achieve IPPS plans with improved production performance.

Petrović et al. (2016a,c) and Miljković & Petrović (2017) provide the details on these manufacturing flexibilities.

In this chapter, the detailed information of alternative resources (specifically on machine tools and cutting tools), as well as sequencing and process flexibilities, are presented in a form of OR network. OR-representation network consists of five types of nodes, i.e., starting node, an intermediate node, ending node, OR node, and JOIN node. The beginning of operation is shown by the starting node and the ending of operation is represented by the ending node while the intermediate nodes have information about manufacturing operations. It consists of the following elements: number of operation, a group of an alternative machine tool, a group of alternative cutting tools, as well as the corresponding processing times according to selected machine tools and cutting tools. OR node defines the start of alternative process plan path while JOIN node states end of the alternative process plan path. Detailed information about flexible process plan networks can be found in Petrović et al. (2016a,c) and Miljković & Petrović (2017).

Figure 9.1 shows three alternative process plan networks for representative parts (job 8, job 10, job 11) adopted from Shin et al. (2010). Taking job 8 as an example, one possible sequence of operations within an alternative process plan is {1, 2, 5, 6, 7, 8, 9, 10}. Furthermore, one alternative process plan of the jobs 10 and 11 can be {7, 8, 9, 10, 11, 14} and {9, 10, 11, 13}, respectively.

On the other hand, each scheduling plan consisting of a sequence of operations for all three jobs is formed according to the information about manufacturing resources (i.e., machine tools and cutting tools) provided by the flexible process plan networks. In the scheduling plan, only one machine tool and one cutting tool can be selected for each operation. For example, for three representative manufacturing parts with flexible process plan networks given in Figure 9.1, a scheduling plan is presented in Figure 9.2. Scheduling plan consists of four strings: scheduling string, process plan string, machine string, and tool string. For more detailed information about parts and string encoding, the reader is referred to Petrović et al. (2016c).

9.3 The Mathematical Modelling of Objective Functions

This section presents the mathematical modelling of fitness functions for optimal IPPS. In order to optimise scheduling plans according to different criteria, three objective functions (objective 1, objective 2, and objective 3) used to obtain optimal scheduling solutions are described hereafter.

The first fitness function is defined by Equation (9.1) in order to minimise the makespan performance measure (objective),

$$object\ 1 = \max\left(c_{ij}\right)\left(c_{ij} \in T_d\left(s_{ij}, c_{ij}\right)\right), \tag{9.1}$$

where c_{ij} is completion time of j-th operation of job i, and s_{ij} is the starting time of j-th operation of job i.

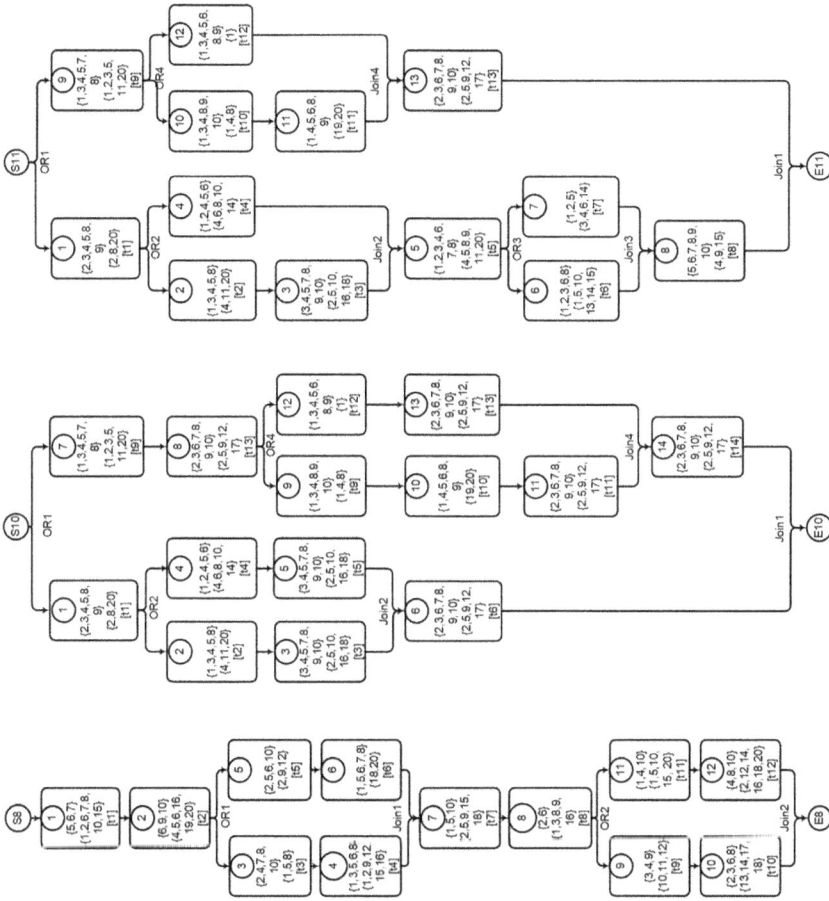

FIGURE 9.1
Flexible process plan networks for job 8, job 10, and job 11.

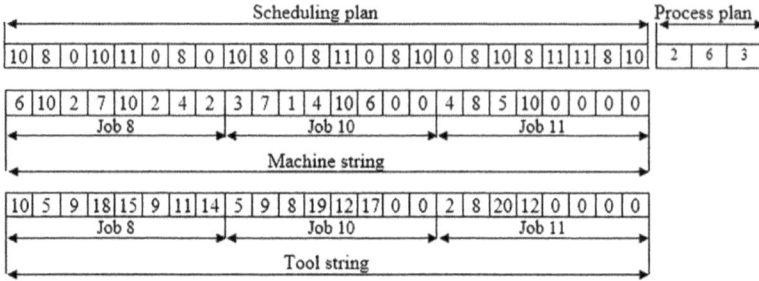

FIGURE 9.2
Encoding of IPPS solution.

The objective of the second fitness function is the maximisation of a balanced level of machine utilisation, which is defined by Equation (9.2):

$$object\ 2 = \min\left(object\ 1 + \sum_{a=1}^{m}\left|\sum p_{ij} - avgmt\right| \right), (o_{ij} \in M_a), \qquad (9.2)$$

where m is the total number of machine tools in a manufacturing environment, $\sum p_{ij}$ is total processing time per machine tool, the average processing time of all machine tools $avgmt$ is calculated as $avgmt = \dfrac{1}{m}\sum_{a=1}^{m}\sum p_{ij}$, o_{ij} is operation j of job i, and M_a is selected machine from set of alternative machine tools for operation o_{ij}.

The objective of the third fitness function is the minimisation of mean flow time, as defined by Equation (9.3):

$$object\ 3 = \min\left(\frac{1}{N}\sum_{i=1}^{n}c_i \right) \qquad (9.3)$$

where c_i is the completion time of job i, and n is the total number of jobs.

9.4 ALO Algorithm

ALO algorithm was originally introduced by Mirjalili (2015). It is a unique and newly established nature-inspired algorithm. This optimisation methodology is inspired by the behaviour of antlions during the process of hunting ants in nature. The two sequential stages in antlions' lifecycles are larva and adult. In the first stage, the antlion larvae (aka "doodlebug") starts the hunting process by making a sand pit after which it patiently waits for prey at the bottom of the pit, Figure 9.3. The sharpness of cone-shaped pit's edge is pronounced and suitable to cause slipping of the prey, as well as causing it to fall

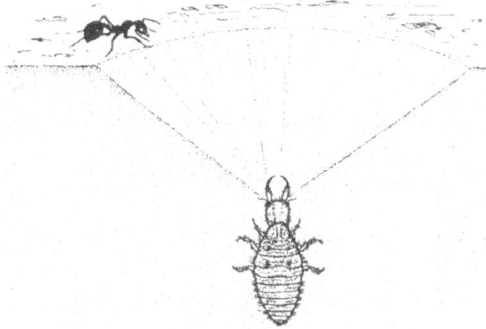

FIGURE 9.3
Position of ant and antlion in cone-shaped pit trap during the hunting process (Link#1).

to the bottom of the pit where it can be easily seied by the antlion. However, in the case that the ant still tries to escape, antlion intelligently throws sand towards the ant and makes it incapable of escaping from the trap. The behaviour of antlions and ants during the hunting (optimisation) process is mathematically described as given hereafter (Mirjalili 2015, Petrović et al. 2016b).

A stochastic movement of ants in nature is the process of food searching (i.e., the random walks), which can be described by Equation (9.4):

$$X(t) = \left[0, cumsum\big(2r(t_1) - 1\big), cumsum\big(2r(t_2) - 1\big), \ldots, cumsum\big(2r(t_n) - 1\big) \right], \quad (9.4)$$

where *cumsum* is the cumulative sum, n represents the maximum iterations number, and $r(t)$ is a stochastic function described by Equation (9.5):

$$r(t) = \begin{cases} 1 \ if \ rand > 0.5 \\ 0 \ if \ rand \leq 0.5 \end{cases}, \quad (9.5)$$

where the random walk step and uniformly distributed random number [0, 1] in the interval are indicated by t and *rand*, respectively.

The ants' positions during the optimisation process are saved in the matrix M_{Ant} (Equation 9.6):

$$M_{Ant} = \begin{bmatrix} A_{1,1} & A_{1,2} & \cdots & A_{1,d} \\ A_{2,1} & A_{2,2} & \cdots & A_{2,d} \\ \vdots & \vdots & \ddots & \vdots \\ A_{n,1} & A_{n,2} & \cdots & A_{n,d} \end{bmatrix}, \quad (9.6)$$

where $A_{i,j}$ indicates the value of variable j of ant i, while n and d are the numbers of ants and variables, respectively.

According to the representation of scheduling plans (see Petrović et al. 2016c), and machine and tool flexibilities considered in this paper, the positions of ants are presented with the matrices M_{Ant_m} and M_{Ant_t} (Equations 9.7 and 9.8):

$$
M_{Ant\_m} = \begin{bmatrix} A_{m1,1} & A_{m1,2} & \cdots & A_{m1,d} \\ A_{m2,1} & A_{m2,2} & \cdots & A_{m2,d} \\ \vdots & \vdots & \ddots & \vdots \\ A_{mn,1} & A_{mn,2} & \cdots & A_{mn,d} \end{bmatrix},
\tag{9.7}
$$

$$
M_{Ant\_t} = \begin{bmatrix} A_{t1,1} & A_{t1,2} & \cdots & A_{t1,d} \\ A_{t2,1} & A_{t2,2} & \cdots & A_{t2,d} \\ \vdots & \vdots & \ddots & \vdots \\ A_{tn,1} & A_{tn,2} & \cdots & A_{tn,d} \end{bmatrix},
\tag{9.8}
$$

where vector $[A_{mn,1}\ A_{mn,2}\ \cdots\ A_{mn,d}]$ presents selected alternative machines for n-th possible scheduling solution, vector $[A_{tn,1}\ A_{tn,2}\ \cdots\ A_{tn,d}]$ presents selected alternative tools for n-th possible scheduling solution, n is the number of ants in a population where each ant represents one scheduling solution (i.e., size of the population) and d is the number of jobs. For example, according to IPPS solution given in Figure 9.2, vector $A_{mn,1}$ is defined as [6 10 2 7 10 2 4 2], where $A_{mn,1}$ (2) presents selected machine 10 for the second operation of the first job. Vector $A_{tn,2}$ is defined as [5 9 8 19 12 17 0 0], where $A_{tn,2}$ (4) presents selected tool 19 for the fourth operation of the second job.

Matrix M_{OA} stores an objective function value of each ant during the optimisation (Equation 9.9):

$$
M_{OA} = \begin{bmatrix} f\left([A_{1,1}, A_{1,2}, \ldots, A_{1,d}]\right) \\ f\left([A_{2,1}, A_{2,2}, \ldots, A_{2,d}]\right) \\ \vdots \\ f\left([A_{n,1}, A_{n,2}, \ldots, A_{n,d}]\right) \end{bmatrix}.
\tag{9.9}
$$

To save positions of antlions, the following matrix $M_{Antlion}$ is used (refer Equation 9.10):

$$
M_{Antlion} = \begin{bmatrix} AL_{1,1} & AL_{1,2} & \cdots & AL_{1,d} \\ AL_{2,1} & AL_{2,2} & \cdots & AL_{2,d} \\ \vdots & \vdots & \ddots & \vdots \\ AL_{n,1} & AL_{n,2} & \cdots & AL_{n,d} \end{bmatrix},
\tag{9.10}
$$

where $AL_{i,j}$ presents the value of variable j of antlion i, and n and d are the numbers of antlions and variables, respectively.

According to methodology for IPPS problem and representation of scheduling plans (see Figure 9.2), positions of the antlions are presented by Equations (9.11) and (9.12):

$$M_{Antlion\_m} = \begin{bmatrix} AL_{m1,1} & AL_{m1,2} & \cdots & AL_{m1,d} \\ AL_{m2,1} & AL_{m2,2} & \cdots & AL_{m2,d} \\ \vdots & \vdots & \ddots & \vdots \\ AL_{mn,1} & AL_{mn,2} & \cdots & AL_{mn,d} \end{bmatrix}, \tag{9.11}$$

$$M_{Antlion\_t} = \begin{bmatrix} AL_{t1,1} & AL_{t1,2} & \cdots & AL_{t1,d} \\ AL_{t2,1} & AL_{t2,2} & \cdots & AL_{t2,d} \\ \vdots & \vdots & \ddots & \vdots \\ AL_{tn,1} & AL_{tn,2} & \cdots & AL_{tn,d} \end{bmatrix}. \tag{9.12}$$

Matrix M_{OAL} defines the objective function value of each antlion during the optimisation (Equation 9.13):

$$M_{OAL} = \begin{bmatrix} f\left(\left[AL_{1,1}, AL_{1,2}, \ldots, AL_{1,d}\right]\right) \\ f\left(\left[AL_{2,1}, AL_{2,2}, \ldots, AL_{2,d}\right]\right) \\ \vdots \\ f\left(\left[AL_{n,1}, AL_{n,2}, \ldots, AL_{n,d}\right]\right) \end{bmatrix}. \tag{9.13}$$

In order to update the ants' positions and keep their movement within the search space, random walks of ants are normalised according to Equation (9.14):

$$X_i^t = \frac{\left(X_i^t - a_i\right) \times \left(d_i - c_i^t\right)}{\left(d_i^t - a_i\right)} + c_i, \tag{9.14}$$

where a_i and b_i are minimum and maximum values of random walk of variable i, respectively, and c_i^t and d_i^t are minimum and maximum values of variable i at iteration t, respectively.

Trapping of ants in the antlion's pits is mathematically modelled by the following equation (9.15):

$$c_i^t = Antlion_j^t + c^t, \; d_i^t = Antlion_j^t + d^t, \tag{9.15}$$

where c^t and d^t are minimum and maximum values of all variables at iteration t, respectively, while $Antlion_j^t$ defines the position of the selected antlion j at iteration t.

Hunting capability of the antlion is modelled by fitness proportional roulette wheel selection. On the other side, the sliding down of the trapped ant towards the antlion located at the bottom of the pit is mathematically modelled by Equation (9.16):

$$c^t = \frac{c^t}{I}, d^t = \frac{d^t}{I}, I = 10^w \cdot \frac{t}{T},$$
(9.16)

where I is a ratio, and w is the constant which is defined according to the current iteration t and the maximum number of iterations T ($w = 2$ if $t > 0.1T$, $w = 3$ if $t > 0.5T$, $w = 4$ if $t > 0.75T$, $w = 5$ if $t > 0.9T$, $w = 6$ if $t > 0.95T$).

Finally, the elitism of the ALO algorithm is introduced to maintain the optimal solution. The antlion with best fitness function obtained so far is considered to be elite. Position of ant i in iteration t $\left(Ant_i^t\right)$ is calculated according to Equation (9.17):

$$Ant_i^t = \frac{R_A^t + R_E^t}{2},$$
(9.17)

where R_A^t indicates the random walk of an ant towards a randomly selected antlion at iteration t, while R_E^t represents the random walk of an ant towards the elite antlion at iteration t.

9.5 Implementation Procedure of ALO Algorithm for IPPS Optimisation

In Table 9.1, the implementation details regarding optimisation of the IPPS problem based on ALO approach are described by the pseudo code.

9.6 Experimental Verification

In the present work, two experimental studies are conducted with regards to the different flexibility levels and types to show the performance of the proposed IPPS methodology. The performance of the ALO algorithm has been assessed and proved with respect to the results produced by the PSO and GA. After preliminary tuning experiments, parameters setting of algorithms for all the experiments are described hereafter.

The GA parameters are set as follows: the maximum number of generations and the size of the population are set to 100, while the mutation probability is 0.10 and the crossover probability is 0.60. Furthermore, parameters of the PSO algorithm are set as follows: both, the maximum number of generations and the population size are set to 100, the inertia weight W is set to linearly

TABLE 9.1

The ALO Algorithm for Optimisation of IPPS

1:	Initialise algorithm parameters (maximum number of iterations, population size, number of variables);
2:	Initialise the first population of ants and antlions randomly using the encoding procedure described in Figure 9.2;
3:	Evaluate the objective function value of each antlion using Equations (9.1)–(9.3);
4:	**Repeat**
5:	Find antlion with best objective function and save it as the elite;
6:	Repeat
7:	Randomly select an antlion for every ant based on fitness proportional roulette wheel mechanism;
8:	Update minimum and maximum values of all variables by Equation (9.16);
9:	Normalise the random walk of ants by using Equation (9.14);
10:	Update positions of an ant by using Equation (9.17);
11:	Apply the procedure of rounding off non-integer solutions to the nearest integer numbers which represent selected machine tool and tool from alternative solution sets;
12:	Until all ants' positions are not generated;
13:	Compute the objective function value of all ants by Equations (9.1)–(9.3);
14:	Replace an antlion's position with the position of the ant if the ant is seized by antlion;
15:	Update elite antlion if it becomes fitter than the elite;
16:	Until the maximum of iterations for IPPS is not reached;
17:	Output elite solution (the optimal scheduling plan which contains a sequence of operations with selected alternative machines and tools).

decrease from 1.2 to 0.4, and the acceleration constants C_1 and C_2 are both set to 2.0. Finally, ALO algorithm's parameters are set as follows: the size of the ants' population and the size of the antlions' population are set to 100. The best and average results are reported after replication of each experiment for 10 times. The stopping criteria to terminate the optimisation process of all algorithms is defined when the maximum number of iterations is reached. MATLAB® software package is used for the implementation and development of all algorithms for experimental and performance comparisons.

9.6.1 Experiment 1

Experiment 1 is designed to study the performance of ALO, PSO, and GA on problems with different levels of flexibility and four types of flexibility (process, machine, sequencing, and tool flexibility). Information about alternative manufacturing resources and corresponding processing times can be found in Shin et al. (2010). The transportation times between the machines are adopted from Petrović et al. (2016c) and presented in Table 9.2. The 24 test-bed problems having varying levels of manufacturing flexibility are formed by the mix of 14 jobs (see experiment 2 from Petrović et al. 2016c), where jobs' combinations for each problem are presented in Table 9.3. Optimal IPPS

TABLE 9.2

Transport Time Between Machine Tools [s] (Petrović et al. 2016c)

Machine No.	1	2	3	4	5	6	7	8	9	10
1	0	4	8	10	12	5	6	14	13	18
2	4	0	3	7	11	5	4	6	10	13
3	8	3	0	5	7	9	8	4	6	10
4	10	7	5	0	4	14	12	6	5	6
5	12	11	7	4	0	18	12	10	6	4
6	5	5	9	14	18	0	6	8	12	15
7	6	4	8	12	12	6	0	3	7	10
8	14	6	4	6	10	8	3	0	4	8
9	13	10	6	5	6	12	7	4	0	4
10	18	13	10	6	4	15	10	8	4	0

TABLE 9.3

Test-Bed Problems of Experiment 1

Problem No.	Number of Jobs	Number of Operations	Job Number
1	6	99	1-2-3-10-11-14
2	6	106	4-5-6-11-14-15
3	6	95	7-8-9-14-15-16
4	6	94	1-4-7-10-14-16
5	6	93	2-5-8-11-14-16
6	6	107	3-6-9-11-15-16
7	6	99	1-4-8-14-15-16
8	6	100	2-6-7-10-14-16
9	6	100	3-5-9-11-14-16
10	9	154	1-2-3-5-6-10-11-14-15
11	9	142	4-7-8-9-10-11-14-15-16
12	9	134	1-4-5-7-8-10-11-14-16
13	9	159	2-3-6-9-10-11-14-15-16
14	9	143	1-2-4-7-8-11-14-15-16
15	9	154	3-5-6-9-10-11-14-15-16
16	11	189	1-2-3-4-5-6-10-11-14-15-16
17	11	179	4-5-6-7-8-9-10-11-14-15-16
18	11	181	1-2-4-5-7-9-10-11-14-15-16
19	11	186	2-3-5-6-8-9-10-11-14-15-16
20	11	179	1-2-4-6-7-8-10-11-14-15-16
21	11	185	2-3-5-6-7-9-10-11-14-15-16
22	11	188	2-3-4-5-6-8-9-10-11-14-16
23	12	195	1-4-5-6-7-8-9-10-11-14-15-16
24	14	233	1-2-3-4-5-6-7-8-9-10-11-14-15-16

solutions are obtained according to the following fitness functions: makespan (object 1, Equation 9.1) and mean flow time (object 3, Equation 9.3). Tool change time for this experiment is 6 s. The solutions of ALO have been compared with the PSO and GA. The best and average fitness values of scheduling plan solutions obtained under object 1 are given in Table 9.4, and those under object 2 are given in Table 9.5. By analyzing the results reported in Table 9.4, it can be pointed out that the ALO algorithm has achieved the best results under object 1 (makespan) for 7 problems. Also, experimental results given in Table 9.5 indicate that the best results obtained by the ALO algorithm for object 3 (mean flow time) are better for 12 problems, while average values are lower for 18 out of 24 problems. Convergence curves and Gantt charts are given in Figures 9.4 and 9.5 for object 1 and 3, respectively.

TABLE 9.4

The Comparative Experimental Results for Best and Average Values of Fitness Function Object 1 (Makespan)

	ALO		PSO		GA	
Problem	Best	Ave	Best	Ave	Best	Ave
1	108	122.5	**103**	113.2	110	115.7
2	117	137.2	113	127.7	**110**	128.8
3	213	232.7	186	207.7	**185**	204.5
4	106	116.3	**96**	104.3	98	109.4
5	204	226.4	187	202.3	**178**	192
6	116	137.2	123	135.4	**99**	121
7	208	233.1	195	220.7	**190**	202.4
8	93	107.1	77	88.60	**80**	93.70
9	94	117.3	92	109.4	**72**	96.20
10	**134**	168.4	154	176.3	146	163.1
11	237	252.2	**212**	238.2	216	235.5
12	215	240	**198**	224.4	213	223.4
13	**141**	164.1	147	161.2	**141**	158.2
14	229	249.3	212	225.4	**190**	220
15	**129**	164.8	146	162.9	140	167.1
16	164	187.7	**159**	190.4	175	187.1
17	231	257.9	**217**	239	233	264
18	**148**	187.6	171	189.3	166	186.2
19	**230**	250.5	**230**	244.1	235	261.2
20	215	259.2	**204**	248	236	252
21	**153**	189.6	169	182.5	163	176
22	232	247.6	**224**	243.6	**224**	247.9
23	248	259.9	230	257.6	**227**	268.1
24	**251**	283.6	254	274.3	262	294.3

The bold represents the minimum value of makesapn for each problem.

TABLE 9.5

The Comparative Experimental Results for Best and Average Values of Fitness Function Object 3 (Mean Flow Time)

	ALO		PSO		GA	
Problem	Best	Ave	Best	Ave	Best	Ave
1	**74.83**	86.17	78.50	85.55	77.50	90.63
2	**68**	89.37	76.83	88.85	85	89.98
3	96.33	107.2	**95.67**	110.88	99	108.78
4	75.50	81.93	**68**	76.55	76.50	79.33
5	**88.67**	102.68	95.170	102.85	100.67	105.80
6	86.17	96.33	**84.83**	97.10	85.670	98.430
7	119.33	130.17	120.83	132.28	**115.67**	124.63
8	64.83	72.37	**60.33**	65.60	63.17	66.62
9	65.83	71.62	67.17	76.42	**63.17**	76.02
10	105.22	119.19	**97.89**	114.02	104.11	117.67
11	**109.44**	130.47	125.22	134.47	123.22	135.27
12	111.33	119.40	**106.44**	121.86	114.78	126.44
13	**97.89**	111.37	98.56	117.50	103.44	114.13
14	**112.89**	126.22	124.67	136.28	114.78	129.02
15	98.89	105.46	**93.78**	115.08	104.11	110.92
16	**108.36**	122.44	116.27	132.88	130.91	143.98
17	**114.45**	134.81	135.45	144.96	138.09	147.50
18	108.18	120.51	**99.36**	122.45	122.82	132.80
19	138.09	142.55	**137.36**	153.28	149.64	164.12
20	**129.82**	139.88	136.82	152.85	136.18	149.94
21	**98.730**	109.46	102.73	119.12	110.55	126.51
22	124.09	138.93	**121.91**	145.65	140.82	153.42
23	**127.67**	142.42	143.75	165.19	148.25	161.49
24	**147.79**	157.55	149.71	169.79	158.64	182.81

The bold represents the minimum value of mean flow time for each problem.

9.6.2 Experiment 2

This experiment is planned to assess the performance of developed algorithms considering the aforementioned three types of manufacturing flexibility on IPPS test-bed problems. Therefore, information about six jobs which manufacturing operations are performed on eight alternative machine tools has been taken from Shao et al. (2009). Also, the transportation times are referred from Shao et al. (2009) and presented in Table 9.6. The objective of this experiment is to obtain optimal IPPS solutions according to minimum makespan (object 1), maximum machine utilisation level (object 2), and minimum mean flow time (object 3). ALO algorithm has been compared with PSO and GA in terms of their performance improvement and given in Table 9.7. Convergence curves, as well as the scheduling results obtained

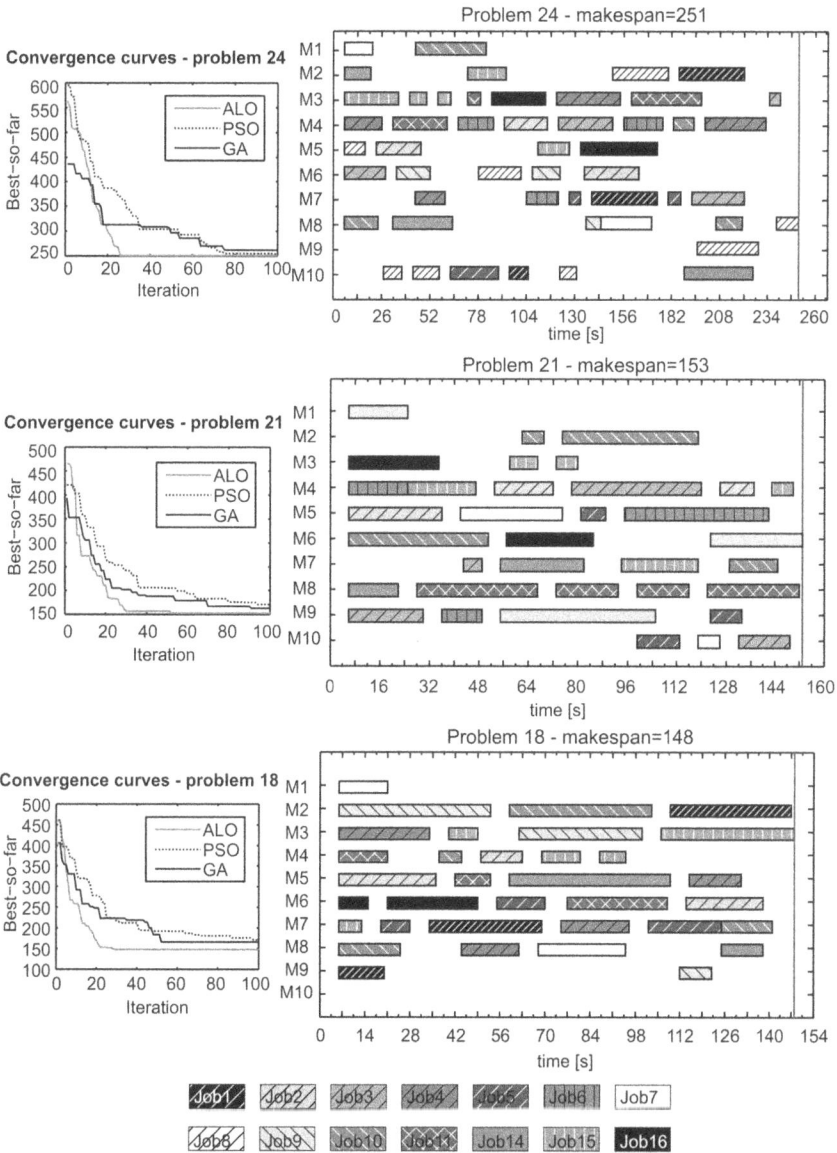

FIGURE 9.4
Convergence curves and Gantt charts achieved by ALO algorithm for fitness function object 1 (makespan).

under all three fitness functions, are presented by Gantt charts in Figure 9.6. Analysis of the results presented in Table 9.7 and Figure 9.6 shows that ALO algorithm achieves the best IPPS solutions for makespan (object 1 = 148) and balanced machine utilisation (object 2 = 187.75), while PSO gives the best IPPS solution for mean flow time (object 3 = 124.5).

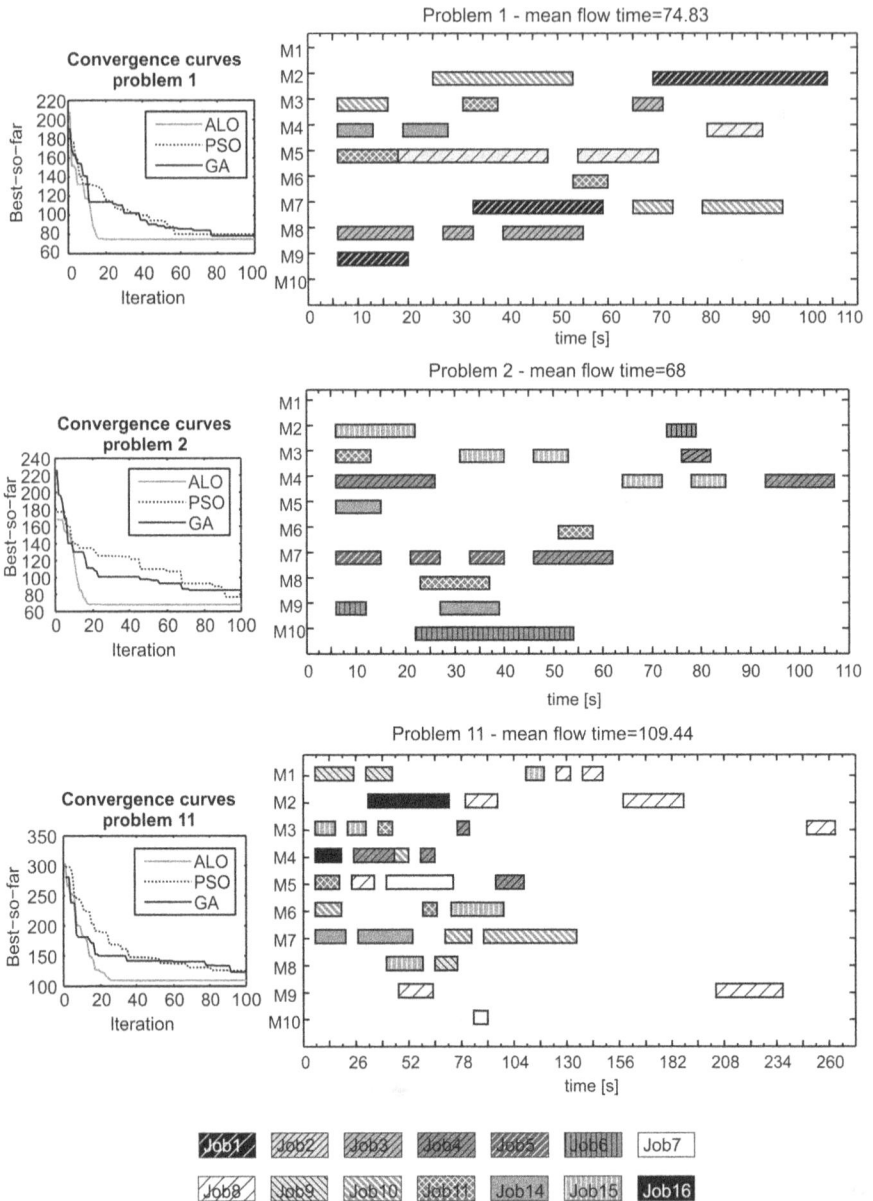

FIGURE 9.5

Convergence curves and Gantt charts achieved by ALO algorithm for fitness function object 3 (mean flow time).

TABLE 9.6

Transport Time Between Machine Tools [s] (Shao et al. 2009)

Machine No.	1	2	3	4	5	6	7	8
1	0	3	7	10	3	5	8	12
2	3	0	4	7	5	3	5	8
3	7	4	0	3	8	5	3	5
4	10	7	3	0	10	8	5	3
5	3	5	8	10	0	3	7	10
6	5	3	5	8	3	0	4	7
7	8	5	3	5	7	4	0	3
8	12	8	5	3	10	7	3	0

TABLE 9.7

The Comparative Results for Best and Average Best Values Obtained Using ALO, GA, and PSO Algorithm

Algorithms		Objectives		
		Object 1	Object 2	Object 3
ALO	Best	**148**	**187.75**	129
	Ave	158.2	212.65	130.7
PSO	Best	**148**	188.75	**124.5**
	Ave	152.9	203.65	127.55
GA	Best	155	204.5	131.5
	Ave	159.2	234.48	133.63

The bold represents the best value of each objective.

9.7 Conclusion

This chapter presents a nature-inspired optimisation methodology used to solve a complex combinatorial NP-hard optimisation problem named IPPS. The suggested ALO methodology is beneficial because it simultaneously selects the process plans and according to them determines the optimal schedules. The representation method based on the OR network is used to simultaneously depict the following four flexibility types of IPPS problem: machine flexibility, process flexibility, sequence flexibility, and tool flexibility. The optimal IPPS solutions are obtained according to three objective functions: (i) minimisation of makespan, (ii) maximisation of the balanced level of machine utilisation, and (iii) minimisation of mean flow time, whose mathematical modelling is also presented in the paper. IPPS solutions are encoded into agents of ALO methodology (ants and antlions) based on information about manufacturing resources represented in a form of OR network.

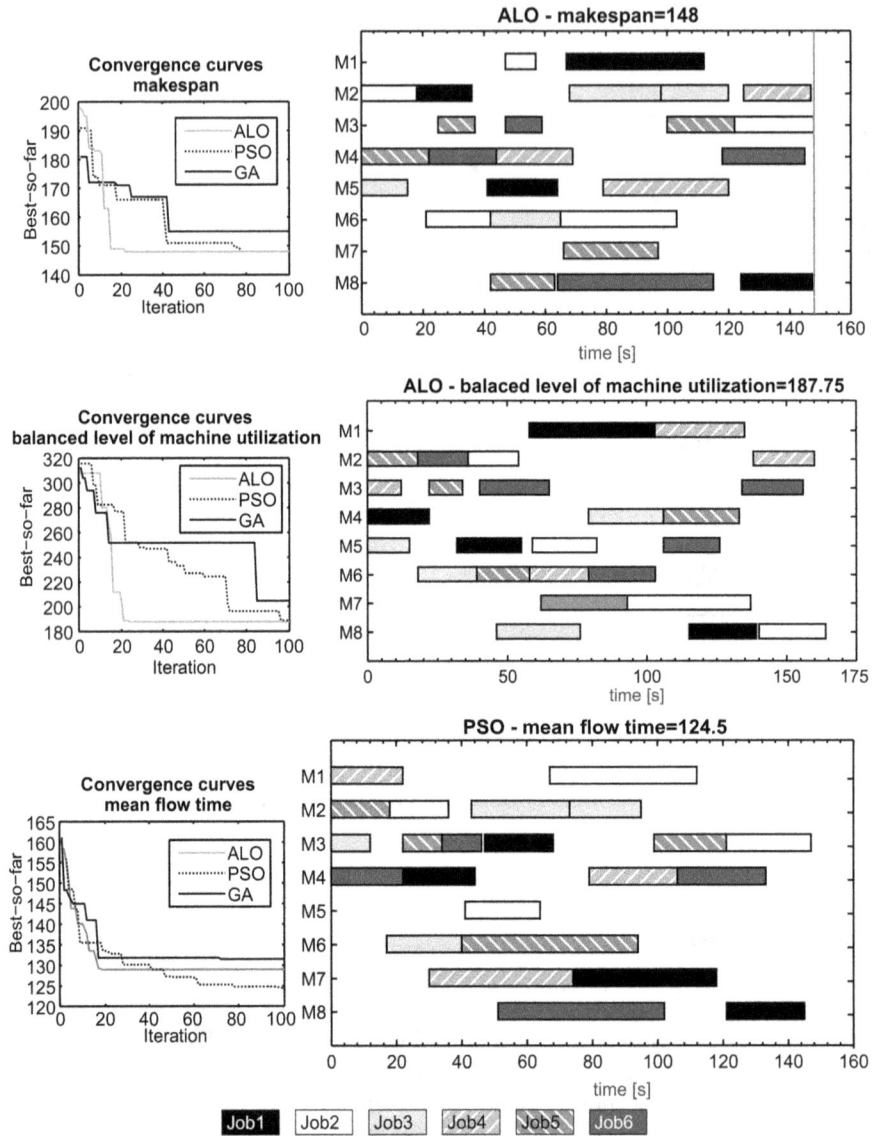

FIGURE 9.6
Convergence curves of ALO, PSO, and GA for all fitness functions (left); Gantt charts of best scheduling plan solution for all fitness functions (right).

The proposed ALO methodology is coded in MATLAB software package and its performance is evaluated through two experimental studies conducted to test algorithms on problems with different types and flexibility levels. Also, the performance of the ALO has been compared with the outcomes of PSO and GA. From the experiments, it can be clearly observed that the ALO algorithm achieves promising results in optimisation of the IPPS problem. In the IPPS, the further scope of research can be the implementation of nature-inspired algorithms in order to optimise the multi-objective IPPS problem.

Acknowledgement

We are grateful for financial aid granted by the "Serbian Ministry of Education, Science and Technological Development" under the title *An innovative, ecologically based approach to the implementation of intelligent manufacturing systems for the production of sheet metal parts*, and grant number TR-35004 (2011–2019).

References

Babić, B., Miljković, Z., Vuković, N., & Antić, V. (2012). Towards implementation and autonomous navigation of an intelligent automated guided vehicle in material handling systems. *Iranian Journal of Science and Technology-Transactions of Mechanical Engineering, 36*(M1), 25–40.

Barzanji, R., Naderi, B., & Begen, M. A. (2019). Decomposition algorithms for the integrated process planning and scheduling problem. *Omega*. doi: 10.1016/j.omega.2019.01.003

Baykasoğlu, A., & Özbakır, L. (2009). A grammatical optimization approach for integrated process planning and scheduling. *Journal of Intelligent Manufacturing, 20*(2), 211–221.

Dai, M., Tang, D., Xu, Y., & Li, W. (2015). Energy-aware integrated process planning and scheduling for job shops. *Proceedings of the Institution of Mechanical Engineers, Part B: Journal of Engineering Manufacture, 229*(1_suppl), 13–26.

Diryag, A., Mitić, M., & Miljković, Z. (2014). Neural networks for prediction of robot failures. *Proceedings of the Institution of Mechanical Engineers, Part C: Journal of Mechanical Engineering Science, 228*(8), 1444–1458.

Guo, Y. W., Li, W. D., Mileham, A. R., & Owen, G. W. (2009). Applications of particle swarm optimisation in integrated process planning and scheduling. *Robotics and Computer-Integrated Manufacturing, 25*(2), 280–288.

Kim, Y. K., Park, K., & Ko, J. (2003). A symbiotic evolutionary algorithm for the integration of process planning and job shop scheduling. *Computers & Operations Research, 30*(8), 1151–1171.

Lee, H., & Kim, S. S. (2001). Integration of process planning and scheduling using simulation based genetic algorithms. *The International Journal of Advanced Manufacturing Technology, 18*(8), 586–590.

Lee, H. C., & Ha, C. (2019). Sustainable integrated process planning and scheduling optimization using a genetic algorithm with an integrated chromosome representation. *Sustainability, 11*(2), 502.

Li, W. D., & McMahon, C. A. (2007). A simulated annealing-based optimization approach for integrated process planning and scheduling. *International Journal of Computer Integrated Manufacturing, 20*(1), 80–95.

Li, X., Zhang, C., Gao, L., Li, W., & Shao, X. (2010). An agent-based approach for integrated process planning and scheduling. *Expert Systems with Applications, 37*(2), 1256–1264.

Li, X., Gao, L., & Li, W. (2012). Application of game theory based hybrid algorithm for multi-objective integrated process planning and scheduling. *Expert Systems with Applications, 39*(1), 288–297.

Li, X., Gao, L., Wang, W., Wang, C., & Wen, L. (2019). Particle swarm optimization hybridized with genetic algorithm for uncertain integrated process planning and scheduling with interval processing time. *Computers & Industrial Engineering.* doi: 10.1016/j.cie.2019.04.028

Lian, K., Zhang, C., Gao, L., & Li, X. (2012). Integrated process planning and scheduling using an imperialist competitive algorithm. *International Journal of Production Research, 50*(15), 4326–4343.

Lihong, Q., & Shengping, L. (2012). An improved genetic algorithm for integrated process planning and scheduling. *The International Journal of Advanced Manufacturing Technology, 58*(5–8), 727–740.

Link#1: http://oldblockwriter.blogspot.com/2012/02/ant-lions-and-angle-of-rest.html, Accessed on April 30, 2019.

Miljković, Z., Mitić, M., Lazarević, M., & Babić, B. (2013a). Neural network reinforcement learning for visual control of robot manipulators. *Expert Systems with Applications, 40*(5), 1721–1736.

Miljković, Z., Vuković, N., Mitić, M., & Babić, B. (2013b). New hybrid vision-based control approach for automated guided vehicles. *The International Journal of Advanced Manufacturing Technology, 66*(1–4), 231–249.

Miljković, Z., & Petrović, M. (2017). Application of modified multi-objective particle swarm optimisation algorithm for flexible process planning problem. *International Journal of Computer Integrated Manufacturing, 30*(2–3), 271–291.

Mirjalili, S. (2015). The ant lion optimizer. *Advances in Engineering Software, 83*, 80–98.

Mitić, M., & Miljković, Z. (2014). Neural network learning from demonstration and epipolar geometry for visual control of a nonholonomic mobile robot. *Soft Computing, 18*(5), 1011–1025.

Mitić, M., & Miljković, Z. (2015). Bio-inspired approach to learning robot motion trajectories and visual control commands. *Expert Systems with Applications, 42*(5), 2624–2637.

Mitić, M., Vuković, N., Petrović, M., & Miljković, Z. (2015). Chaotic fruit fly optimization algorithm. *Knowledge-Based Systems, 89*, 446–458.

Mitić, M., Vuković, N., Petrović, M., & Miljković, Z. (2018). Chaotic metaheuristic algorithms for learning and reproduction of robot motion trajectories. *Neural Computing and Applications, 30*(4), 1065–1083.

Morad, N., & Zalzala, A. M. S. (1999). Genetic algorithms in integrated process planning and scheduling. *Journal of Intelligent Manufacturing, 10*(2), 169–179.

Petrović, M., Miljković, Z., Babić, B., Vuković, N., & Čović, N. (2012). Towards a conceptual design of intelligent material transport using artificial intelligence. *Strojarstvo, 54*(3), 205–219.

Petrović, M., Miljković, Z., & Babić, B. (2013). Integration of process planning, scheduling, and mobile robot navigation based on TRIZ and multi-agent methodology. *FME Transactions, 41*(2), 120–129.

Petrović, M., Petronijević, J., Mitić, M., Vuković, N., Plemić, A., Miljković, Z., & Babić, B. (2015). The ant lion optimization algorithm for flexible process planning. *Journal of Production Engineering, 18*(2), 65–68.

Petrović, M., Mitić, M., Vuković, N., & Miljković, Z. (2016a). Chaotic particle swarm optimization algorithm for flexible process planning. *The International Journal of Advanced Manufacturing Technology, 85*(9–12), 2535–2555.

Petrović, M., Petronijević, J., Mitić, M., Vuković, N., Miljković, Z., & Babić, B. (2016b). The ant lion optimization algorithm for integrated process planning and scheduling. *Applied Mechanics and Materials, 834*, 187–192. Trans Tech Publications.

Petrović, M., Vuković, N., Mitić, M., & Miljković, Z. (2016c). Integration of process planning and scheduling using chaotic particle swarm optimization algorithm. *Expert Systems with Applications, 64*, 569–588.

Phanden, R. K., Jain, A., & Verma, R. (2013). An approach for integration of process planning and scheduling. *International Journal of Computer Integrated Manufacturing, 26*(4), 284–302.

Seker, A., Erol, S., & Botsali, R. (2013). A neuro-fuzzy model for a new hybrid integrated process planning and scheduling system. *Expert Systems with Applications, 40*(13), 5341–5351.

Shao, X., Li, X., Gao, L., & Zhang, C. (2009). Integration of process planning and scheduling—a modified genetic algorithm-based approach. *Computers & Operations Research, 36*(6), 2082–2096.

Shin, K. S., Park, J. O., & Kim, Y. K. (2010). Test-bed problems for multi-objective FMS process planning using multi-objective symbiotic evolutionary algorithm. *Computers & Operations Research 38*(3), 702–712.

Srinivas, P. S., RamachandraRaju, V., & Rao, C. S. P. (2012). Optimization of process planning and scheduling using ACO and PSO algorithms. *International Journal of Emerging Technology and Advanced Engineering, 2*(10), 343–354.

Vuković, N., Mitić, M., & Miljković, Z. (2015). Trajectory learning and reproduction for differential drive mobile robots based on gmm/hmm and dynamic time warping using learning from demonstration framework. *Engineering Applications of Artificial Intelligence, 45*, 388–404.

Vuković, N., Petrović, M., & Miljković, Z. (2018). A comprehensive experimental evaluation of orthogonal polynomial expanded random vector functional link neural networks for regression. *Applied Soft Computing, 70*, 1083–1096.

Wong, T. N., Leung, C. W., Mak, K. L., & Fung, R. Y. K. (2006a). An agent-based negotiation approach to integrate process planning and scheduling. *International Journal of Production Research, 44*(7), 1331–1351.

Wong, T. N., Leung, C. W., Mak, K. L., & Fung, R. Y. K. (2006b). Integrated process planning and scheduling/rescheduling—an agent-based approach. *International Journal of Production Research, 44*(18–19), 3627–3655.

Wong, T. N., Zhang, S., Wang, G., & Zhang, L. (2012). Integrated process planning and scheduling–multi-agent system with two-stage ant colony optimisation algorithm. *International Journal of Production Research, 50*(21), 6188–6201.

Yu, M., Zhang, Y., Chen, K., & Zhang, D. (2015). Integration of process planning and scheduling using a hybrid GA/PSO algorithm. *The International Journal of Advanced Manufacturing Technology, 78*(1–4), 583–592.

Zhang, X. D., & Yan, H. S. (2005). Integrated optimization of production planning and scheduling for a kind of job-shop. *The International Journal of Advanced Manufacturing Technology, 26*(7–8), 876–886.

Zhang, Z., Tang, R., Peng, T., Tao, L., & Jia, S. (2016). A method for minimizing the energy consumption of machining system: integration of process planning and scheduling. *Journal of Cleaner Production, 137,* 1647–1662.

Zhang, S., & Wong, T. N. (2018). Integrated process planning and scheduling: an enhanced ant colony optimization heuristic with parameter tuning. *Journal of Intelligent Manufacturing, 29*(3), 585–601.

10

A Review on Testbed Problems for Integration of Process Planning and Scheduling

Rakesh Kumar Phanden

Amity University Uttar Pradesh

Halil Ibrahim Demir

Sakarya University

Ajai Jain

National Institute of Technology Kurukshetra

CONTENTS

10.1 Introduction 207
10.2 Number of Jobs and Machines in IPPS Testbed Problems 208
10.3 Manufacturing Flexibility in IPPS Testbed Problems 210
10.4 Review of IPPS Testbed Problems 218
10.5 Conclusion 220
References 221

10.1 Introduction

Numerous researchers have created hypothetical testbed problems to test their proposed Integration of Process Planning and Scheduling (IPPS) models. These testbed problems are considered as a benchmark to compare the performance of their proposed IPPS approaches. However, the questing of a suitable testbed problem is a cumbersome task for the beginners in IPPS research. Therefore, this chapter presents a review of IPPS testbed problems available in the literature. Consequently, a beginner will be able to conveniently access most of the popular IPPS testbed problems at one place instead of exploring full literature to test and prove the performance of their proposed approach with existing IPPS approaches. Another motivation to conduct this review arises from the relationship between the size of the IPPS problem (with respect to the number of jobs and machine and/or manufacturing flexibility) and computational time of proposed IPPS algorithms. Various researchers have proved that the running time of IPPS model increases exponentially with respect to the problem size and vice versa (Kim and Egbelu 1998; Haddadzade et al. 2014). Therefore, the selection of the size of the IPPS testbed problem is a learned decision in order to prove the effectiveness of the developed IPPS model.

The process planning and scheduling functions involve the assignment of resources and both are interrelated in nature. Also, both functions focus on single criterion optimisation to find optimal production schedule and process plans. Therefore, some researchers have considered the job shop scheduling problems to test and implement in IPPS environment (Li et al. 2010a,b). In fact, they have explored the manufacturing flexibility available at the process planning level as per NLA and solved the job shop scheduling problem in the context of IPPS performance measure. However, a real manufacturing system involves more than one optimisation criterion.

10.2 Number of Jobs and Machines in IPPS Testbed Problems

The job shop system is most suitable for small and medium scale production in batches. It has been observed during the literature review conducted in Chapter 2 that the varying number of machines and jobs were considered by various researchers to solve their formalised IPPS models. In general, the number of machines and jobs are varying from 3 to 30, subject to the job shop configuration (i.e., without consideration of more than one unit production quantity), as shown in Tables 10.1–10.4. Accordingly, the IPPS testbed problem size may be categorised as small size, medium size, and large size if the number of machines and jobs varies from 3 to 10, 11 to 20, and 21 to 30, respectively. Additionally, this categorisation may further be subdivided into small to medium size problems, or medium to large size problems, according to the change in the category of either the number of machines or the number of jobs. However, according to this classification, there is no large size IPPS testbed problem available in which both the machines and the jobs are more than 20 in number. Also, there is no IPPS testbed problem with medium to large size problem. Moreover, numerous researchers have considered more than one production quantity of jobs for a job type, this scenario may be termed as extended testbed problem. In addition, many researchers have considered uniformly or randomly distributed sizes of job shop (Moon et al. 2008). On the other hand, Zhang and Wu (2010) classify the job shop scheduling problem in two categories, i.e., small scale and large scale, in which the number of jobs and the number of machines are varying from 10 to 1000 and 10 to 30, respectively.

TABLE 10.1

Small Size IPPS Testbed Problems

Authors	N	M	Processing Time and Alternative Machines Data	Adopted by	Remarks
Sundaram and Fu (1988)	5	5	Table 10.5	Morad and Zalzala (1999); Moon et al. (2008); Li et al. (2010a); Li et al. (2009); Shao et al. (2009)	—
Nasr and Elsayed (1990)	4	6	Table 10.6	Li et al. (2010a)	—
Chryssolouris et al. (1992)	10	9	Table 10.7	Jain and Elmaraghy (1997); Wong et al. (2006); Lihong and Shengping (2012)	—
Lee and Dicesare (1994)	5	3	Table 10.8	Leung et al. (2010)	—
Morad and Zalzala (1999)	4	3	Table 10.9	Shao et al. (2009)	They have considered 5 units of setup time for each operation
Moon et al. (2008)	5	5	Table 10.10	Shao et al. (2009)	AND–OR network based
Shao et al. (2009)	6	8	—	—	Transportation time between machines has been considered as shows in Table 11.8 Operational flexibility, processing flexibility and sequencing flexibility has been considered
Li et al. (2010a)	6	5	Table 10.11	—	Only three alternative process plans of each part type are given
Li et al. (2010a)	6	5	Table 10.12	—	Alternative operation sequences considering precedence constraints are given
Haddadzade et al. (2014)	6	8	—	—	AND–OR network based

N, number of jobs; M, number of machines.

TABLE 10.2

Extended Size IPPS Testbed Problems

Authors	N	M	Extended Data	Adopted from	Extended Factors Considered
Morad and Zalzala (1999)	4	3	Tables 10.13–10.15	–	Processing cost, percentage of scrap produced, production quantity, cost of raw material, cost of scrap produced
Moon and Seo (2005)	5	6	Tables 10.16 and 10.17	Shao et al. (2009)	Lot size = 40, 70, 60, 30, and 60. Load size = 10 unit for each part type Setup time is generated randomly between 1 and 50
Li et al. (2009)	5	5	Table 10.18	Sundaram and Fu (1988)	Three production orders
Li et al. (2010a)	5	5	Table 10.19	Sundaram and Fu (1988)	Eight production orders
Li et al. (2010a)	4	6	Table 10.20	Nasr and Elsayed (1990)	Eight production orders

N, number of jobs; M, number of machines.

TABLE 10.3

Small to Medium Size IPPS Testbed Problems

Authors	N	M	Adopted by	Remarks
Jain et al. (2006)	18	4	Li et al. (2010a)	–
Park and Choi (2006)	3	13	–	They have given process sheets of three products with precedence operation constraints
Özgüven et al. (2010)	20	10	–	They have generated the processing time and operation sequences uniformly distributed

N, number of jobs; M, number of machines.

10.3 Manufacturing Flexibility in IPPS Testbed Problems

The manufacturing flexibility is a very complex and multi-dimensional subject. It has been studied in many directions such as; (i) horizontal flexibility (i.e., at up-stream level in design and purchase functions or at down-stream level in distribution and customer services), (ii) vertical flexibility

TABLE 10.4

Classification of IPPS Testbed Problem According to
Manufacturing Flexibility by Kim et al. (2003)

Problem No.	N	M	Levels of Manufacturing Flexibility
1	6	15	Low processing flexibility
2	6	15	Medium processing flexibility
3	6	15	High processing flexibility
4	6	15	Low processing flexibility
5	6	15	Medium processing flexibility
6	6	15	High processing flexibility
7	6	15	Low operation flexibility
8	6	15	Medium operation flexibility
9	6	15	High operation flexibility
10	9	15	Low or medium processing flexibility
11	9	15	Medium or high processing flexibility
12	9	15	Low or medium sequence flexibility
13	9	15	Medium or high sequence flexibility
14	9	15	Low or medium operation flexibility
15	9	15	Medium or high operation flexibility
16	12	15	Low or medium processing flexibility
17	12	15	Medium or high processing flexibility
18	12	15	Low or medium sequence flexibility
19	12	15	Medium or high sequence flexibility
20	12	15	Low or medium operation flexibility
21	12	15	Medium or high operation flexibility
22	15	15	–
23	15	15	–
24	18	15	–

N, number of jobs; M, number of machines.

(i.e., estimated for the single resource of a system at micro level or the whole system at macro level), (iii) temporal flexibility (it is an instantaneous flexibility according to the design adequacy and reasoning in order to select the resources and/or to modify the design features) and, (iv) flexibility by the object of the variation (e.g., short-term flexibility with respect to volumes or mix, medium to long-term flexibility with respect to product and process innovation and for the expansion of manufacturing capacity) (De Toni and Tonchia 1998; Jain et al. 2013). However, in IPPS problem the manufacturing flexibility is assessed during process planning (from operation/route sheet), in which multiple process plans for each job type are created (or networked) with "Operation Flexibility" (OF) (i.e., the possibility of performing an operation on more than one machine), "Sequencing Flexibility" (SF) (i.e., the possibility of interchanging the sequence in which the required

TABLE 10.5

Testbed Problem of Sundaram and Fu (1988)

		Machine No.				
Job No.	Operation No.	1	2	3	4	5
1	1	5	3	–	–	–
	2	–	7	–	–	–
	3	–	–	6	–	–
	4	–	–	–	3	–
2	1	7	–	–	–	–
	2	–	4	6	–	–
	3	–	–	7	7	–
	4	–	–	–	–	10
3	1	4	5	8	–	–
	2	–	–	–	5	–
	3	–	–	–	6	5
	4	–	–	–	–	4
4	1	–	2	6	–	–
	2	–	–	8	–	–
	3	–	–	3	8	–
	4	–	–	–	7	4
5	1	3	–	5	–	–
	2	–	–	7	–	–
	3	–	–	–	9	6
	4	–	–	–	–	3

TABLE 10.6

Testbed Problem of Nasr and Elsayed (1990)

		Machine No.					
Job No.	Operation No.	1	2	3	4	5	6
1	1	2	3	4	–	–	–
	2	–	3	–	2	4	–
	3	1	4	5	–	–	–
2	1	3	–	5	–	2	–
	2	4	3	–	–	6	–
	3	–	–	4	–	7	11
3	1	5	6	–	–	–	–
	2	–	4	–	3	5	–
	3	–	–	13	–	9	12
4	1	9	–	7	9	–	–
	2	–	6	–	4	–	5
	3	1	–	3	–	–	3

TABLE 10.7

Testbed Problem of Chryssolouris et al. (1992)

Job No.	Operation No.	Machines								
		1	2	3	4	5	6	7	8	9
1	1	97	88	95	86	90	–	–	–	–
	2	–	–	–	–	–	6	5	–	–
	3	34	34	33	31	32	–	–	–	–
	4	–	–	–	–	–	–	–	11	–
	5	–	–	–	–	–	–	–	–	7
2	1	35	36	32	32	36	–	–	–	–
	2	–	–	–	–	–	–	–	–	10
	3	65	59	62	60	62	–	–	–	–
	4	–	–	–	–	–	–	–	15	–
3	1	54	48	48	50	51				
4	1	41	43	39	43	38	–	–	–	–
	2	–	–	–	–	–	–	–	–	9
	3	–	–	–	–	–	–	–	8	–
	4	86	93	84	84	88	–	–	–	–
	5	–	–	–	–	–	6	6	–	–
5	1	85	82	89	87	89	–	–	–	–
	2	–	–	–	–	–	–	–	–	6
6	1	60	66	60	65	66	–	–	–	–
	2	–	–	–	–	–	–	–	–	7
	3	–	–	–	–	–	–	–	20	
	4	–	–	–	–	–	6	6	–	–
7	1	78	84	80	86	86	–	–	–	–
	2	–	–	–	–	–	–	–	–	9
	3	–	–	–	–	–	–	–	19	–
	4	66	63	59	63	63	–	–	–	–
	5	–	–	–	–	–	–	–	–	10
8	1	23	24	24	24	23	–	–	–	–
	2	–	–	–	–	–	10	10	–	–
9	1	62	62	71	67	69	–	–	–	–
	2	–	–	–	–	–	8	8	–	–
	3	–	–	–	–	–	–	–	–	10
	4	48	44	45	48	44	–	–	–	–
10	1	61	65	60	59	63	–	–	–	–
	2	–	–	–	–	–	–	–	–	8
	3	–	–	–	–	–	–	–	12	–

manufacturing operations are performed), and "Processing Flexibility" (PF) (i.e., the possibility of producing the same manufacturing feature with alternative operations or sequence of operations) (Benjaafar and Ramakrishnan 1996). In this direction, Kim et al. (2003), Shao et al. (2009), and Haddadzade

TABLE 10.8

Testbed Problem of Lee and Dicesare (1994)

		Machines		
Job No.	Operation No.	1	2	3
1	1	70	–	40
	2	–	30	–
	3	30	–	60
	4	20	40	–
2	1	80	120	–
	2	–	–	40
	3	70	140	–
	4	80	–	40
3	1	100	150	80
	2	–	20	60
	3	20	–	40
	4	60	20	–
4	1	–	90	50
	2	60		20
	3	–	70	120
	4	90	60	30
5	1	10	–	150
	2	–	70	140
	3	50	80	–
	4	40	60	80

TABLE 10.9

Testbed Problem of Morad and Zalzala (1999)

		Machines		
Job No.	Operation No.	1	2	3
1	1	15	10	20
	2	–	15	20
	3	15	–	20
2	1	20	25	25
	2	15	–	10
	3	–	5	10
3	1	15	10	–
	2	15	20	15
	3	15	10	20
4	1	15	–	20
	2	20	25	25
	3	05	–	10

TABLE 10.10

Testbed Problem of Moon et al. (2008)

Job No.	Operation No.	Machines					Alternative Operation Sequences
		1	2	3	4	5	
1	1	5	3	–	–	–	1-2
	2	–	5	–	–	–	
2	1	–	–	6	5	–	1-2
	2	–	–	–	–	4	
3	1	5	4	–	–	–	1-2-3 OR 1-3-2 OR 3-1-2
	2	–		2	3	–	
	3	–	5	–	–	–	
4	1	–	–	4	–	–	1-2
	2	–	–	–	5	–	
5	1	4	3	–	–	–	1-2-3-4 OR 1-3-2-4 OR 3-1-2-4
	2	2	4	–	–	–	
	3	–	–	5	–	–	
	4	–	–	4	–	3	

TABLE 10.11

Testbed Problem of Li et al. (2010a) with Alternative Process Plans

Job No.	Alternative Process Plans {Machine No. (Processing Time)}
1	1(10)-3(15)-2(10)-5(20)-4(10)
	1(10)-3(20)-5(20)-4(15)
	2(10)-4(20)-3(20)-5(15)
2	1(10)-3(18)-4(12)-5(15)
	3(8)-2(12)-1(14)-4(13)-5(8)
	2(10)-4(13)-3(18)-5(14)
3	3(12)-1(12)-5(10)-4(12)
	1(10)-2(8)-3(14)-4(6)-5(10)
	2(6)-1(12)-3(12)-4(8)-5(10)
4	1(6)-3(12)-2(8)-5(12)-4(10)
	3(10)-1(8)-2(9)-4(12)-5(8)
	2(8)-3(12)-1(6)-5(14)-4(8)
5	1(10)-2(15)-4(9)-5(10)
	3(10)-2(16)-4(8)-5(8)
	4(6)-3(10)-2(8)-1(10)-5(8)
6	5(6)-2(16)-3(10)-4(10)
	1(9)-2(7)-4(8)-5(8)-3(9)
	5(6)-1(10)-2(8)-3(8)-4(9)

TABLE 10.12

Testbed Problem of Li et al. (2010a) with Alternative Operation Sequences and Precedence Constrains

Job No.	Operation No.	Machines					Alternative Operation Sequences
		1	2	3	4	5	
1	1	5	5	–	6	–	1-2
	2	–	5	–	–	6	
2	1	–	–	6	–	6	1-2
	2	–	5	–	5	6	
3	1	4	–	3	5	–	1-2-3 OR 1-3-2
	2	8	–	7	–	8	
	3	–	5	–	–	–	
4	1	4	–	4	–	–	1-2-3 OR 1-3-2
	2	–	8	7	–	–	
	3	–	–	–	5	5	
5	1	–	–	4	–	3	1-2-3-4 OR 2-1-3-4 OR 1-3-2-4
	2	7	–	–	7	–	
	3	8	7	–	–	8	
	4	5	–	–	6	–	
6	1	8	7	–	8	–	1-2-3-4 OR 2-1-3-4
	2	–	–	5	5	–	
	3	–	8	–	–	–	
	4	7	–	8	–	8	

TABLE 10.13

Processing Cost for Testbed Problem of Morad and Zalzala (1999)

Job No.	Operation No.	Machines		
		1	2	3
1	1	20	15	10
	2	–	50	15
	3	20	–	25
2	1	50	15	15
	2	15	–	50
	3	–	90	05
3	1	60	15	–
	2	15	15	50
	3	50	15	20
4	1	20	–	25
	2	50	15	15
	3	15	–	20

TABLE 10.14

Percentage of Scrap Produced in Testbed Problem of Morad and Zalzala (1999)

Job No.	Operation No.	Machines 1	2	3
1	1	05	10	15
	2	–	05	10
	3	05	–	10
2	1	05	10	10
	2	05	–	05
	3	–	05	10
3	1	15	10	–
	2	10	10	05
	3	05	10	15
4	1	05	–	10
	2	05	10	10
	3	05	–	05

TABLE 10.15

Quantity and Cost of Raw Material and Scrap in Testbed Problem of Morad and Zalzala (1999)

Job No.	Quantity	Cost of Raw Material (USD/Unit)	Cost of Scrap (USD/Unit)
1	10	10	2
2	20	20	3
3	10	30	4
4	20	40	5

et al. (2014) have proposed AND-OR network–based IPPS testbed problems for the varying levels of PF, SF, and OF. Various other researchers such as Sundaram and Fu (1988), Nasr and Elsayed (1990), Chryssolouris et al. (1992), Lee and Dicesare (1994), Morad and Zalzala (1999) have presented testbed problems with alternative machines and operation precedence constraints in tabular form. In addition, the job shop scheduling environment has been categorized into partial and total manufacturing flexibility problems. In total flexible problems, each operation can be processed on any of the machines in the shop. In partial flexible problems, some operations are only achievable on part of the available machines in the shop (Chaudhry and Khan 2016; Kacem et al. 2002). It can be observed that partially flexible problems are more realistic in nature as compare to the total flexibility problems.

TABLE 10.16

Testbed Problem of Moon and Seo (2005)

Job No.	Operation No.	Machines					
		1	2	3	4	5	6
1	1	7	–	–	5	–	–
	2	7	–	–	6	–	–
	3	–	6	5	–	8	–
	4	6	–	–	–	–	5
2	1	–	9	–	8	–	–
	2	3	5	–		6	–
	3	8	–	12	9		8
3	1	–	–	5		8	
	2	10	–	–	10	–	7
	3	6	5	–	–	6	–
	4	15	–	–	6	–	5
	5	–	6	–	–	5	
4	1	–	–	6	6	–	8
	2	–	5	–	–	9	–
	3	–	–	6	4	–	–
	4	–	5	–	3	–	5
	5	5	–	–	–	4	–
5	1	–	8	–	6	–	8
	2	–	7	10	–	8	
	3	13	–	–	–	8	9
	4	–	–	7	6	–	–

TABLE 10.17

Transportation Time in Testbed Problem of Moon and Seo (2005)

Machines	1	2	3	4	5	6
1	0	5	6	–	–	–
2	5	0	7	–	–	–
3	6	7	0	–	–	–
4	–	–	–	0	5	6
5	–	–	–	5	0	7
6	–	–	–	6	7	0

10.4 Review of IPPS Testbed Problems

Table 10.1 presents the details of small size IPPS testbed problems. These problems are fixed with respect to the number of jobs and machines. Kim and Egbelu (1998) presented IPPS testbed problem with uniformly distributed

TABLE 10.18

Production Quantity of Each Job Type for Extended Testbed Problem of Sundaram and Fu (1988) in Li et al. (2009)

Order No.	No. (or Type) of Jobs	Quantity of each job type				
		Job – 1	Job – 2	Job – 3	Job – 4	Job – 5
1	5	20	20	20	20	20
2	4	25	25	25	25	0
3	5	15	15	20	25	25

TABLE 10.19

Quantity of Each Job Type in Production Order Taken by Li et al. (2010a)

Job No.	Order No							
	1	2	3	4	5	6	7	8
1	10	20	25	15	30	40	60	80
2	10	20	25	15	25	40	70	100
3	10	20	25	20	30	30	50	80
4	10	20	25	25	35	50	50	60
5	10	20	00	25	30	40	70	80

TABLE 10.20

Production Quantity of Each Job Type for Testbed Problem of Nasr and Elsayed (1990) Taken by Li et al. (2010a)

Order No.	Quantity			
	Job – 1	Job – 2	Job – 3	Job – 4
1	2	2	2	2
2	10	8	8	6
3	24	10	20	10
4	25	25	25	25
5	30	30	40	50
6	50	50	60	40
7	80	80	70	70
8	100	100	100	100

processing time between 1 and 10. They have given various combinations with 2–10 numbers of jobs and 3–10 numbers of machines. They have also provided the varying number of alternative process plans from 2 to 5 for each part type. It has been observed that the problem reported by Sundaram and Fu (1988) is more popular among IPPS researchers.

Kim et al. (2003) presented 18 jobs and 15 machines IPPS testbed problems, as shown in Table 10.4. They adopted AND-OR network to characterize the 24 problems with varying number of jobs from 6 to 18 and with three levels of low, medium, and high PF, SF, and OF. This testbed has been adopted by numerous researchers namely Wong et al. (2006), Leung et al. (2010), Li et al. 2010b, Lian et al. (2012), as well as Phanden et al. (2013) in order to assess the performance of the proposed IPPS approach. Moreover, the literature review reveals that there is no large size IPPS testbed problem available in which the number of jobs and number of machines are more than 20. In addition, researchers consider the extended problem as a large problem in which the problem complexity is increased either by considering more than one quantity of jobs in a production order, or by increasing the manufacturing flexibility levels through incorporating the alternative sequences of operations. In this direction, Table 10.2 shows various extended problems, which consider varying sizes of production orders, processing cost, etc. Table 10.3 shows a few problems with small to medium size of IPPS testbed problems. However, the medium to large size IPPS testbed problems are not perceived in literature

10.5 Conclusion

This chapter presents a review on testbed problems adopted in literature to solve IPPS. The trend shows that various researchers have tested the proposed approaches on small scale testbed problems and extended small scale testbed problems. Very few have shown interest to experiment with medium scale testbed problems. A rigorous literature review found that no authors have experimented with large scale IPPS testbed problems. Although, researchers have focused to test on low, medium, and high levels of manufacturing flexibility in terms of processing, operations, and sequencing. In this direction, an important and diversified testbed problem has been presented by Kim et al. (2003). Moreover, literature review reveals that authors have extended small scale problems with variety of production orders that composed of different production quantity, lot sizes, transportation time, and setup time in order to map the experimentation to the proposed IPPS approach. It has been noticed that the IPPS testbed problems having uniform distribution of processing time are not as popular as the problems with fixed (deterministic) processing time and alternative machines.

In addition, this work can be extended to present the optimised solutions obtained for IPPS testbed problems. In literature, the proposed IPPS models have been tested for varying performance measures such as makespan, mean flow time, tardiness, balanced machine utilisation level, and many more. In order to validate the developed IPPS approaches, researchers focused on proving the effectiveness of proposed approaches over the existing approach. Thus, a beginner can better comprehend the trending, latest, and promising IPPS approach which can yield an optimised solution.

References

Benjaafar, S. & Ramakrishnan, R. (1996). Modeling, measurement and evaluation of sequencing flexibility in manufacturing systems. *International Journal of Production Research*, 4(5), 1195–1220.

Chan, F. T. S., Kumar, V., & Tiwari, M. K. (2006). Optimizing the performance of an integrated process planning and scheduling problem: An AIS-FLC based approach. In *Proceedings of the CIS*, Bangkok, Thailand. IEEE.

Chaudhry, I. A., & Khan, A. A. (2016). A research survey: Review of flexible job shop scheduling techniques. *International Transactions in Operational Research*, 23(3), 551–591.

Chryssolouris, G., Pierce, J. E., & Dicke, K. (1992). A decision-making approach to the operation of flexible manufacturing systems. *International Journal of Flexible Manufacturing Systems*, 4(3–4), 309–330.

De Toni, A., & Tonchia, S. (1998). Manufacturing flexibility: A literature review. *International Journal of Production Research*, 36(6), 1587–1617.

Haddadzade, M., Razfar, M. R., & Fazel Zarandi, M. H. (2014). Integration of process planning and job shop scheduling with stochastic processing time. *International Journal of Advanced Manufacturing Technology*, 71(1–4), 241–252.

Jain, A., Jain P. K., & Singh, I. P. (2006). An integrated scheme for process planning and scheduling in FMS. *International Journal of Advanced Manufacturing Technology*, 30, 1111–1118.

Jain, A. K., Elmaraghy, H. A. (1997). Production scheduling/rescheduling in flexible manufacturing. *International Journal of Production Research*, 35, 281–309.

Jain, A., Jain, P. K., Chan, F. T., & Singh, S. (2013). A review on manufacturing flexibility. *International Journal of Production Research*, 51(19), 5946–5970.

Kacem, I., Hammadi, S., & Borne, P. (2002). Approach by localization and multiobjective evolutionary optimization for flexible job-shop scheduling problems. *IEEE Transactions on Systems, Man, and Cybernetics, Part C (Applications and Reviews)*, 32(1), 1–13.

Kim, K. H., & Egbelu, P. J. (1998). A mathematical model for job shop scheduling with multiple process plan consideration per job. *Production Planning & Control*, 9(3), 250–259.

Kim, Y. K., Park, K., & Ko, J. (2003). A symbiotic evolutionary algorithm for the integration of process planning and job shop scheduling. *Computers & Operations Research*, 30(8), 1151–1171.

Lee, D. Y., & Dicesare, F. (1994). FMS scheduling using Petri nets and heuristic search. *IEEE Transactions on Robotics and Automation*, 10(2), 123–132.

Leung, C. W., Wong, T. N., Mak, K. L., & Fung, R. Y. K. (2010). Integrated process planning and scheduling by an agent-based ant colony optimization. *Computers and Industrial Engineering*, 59(1), 166–180. doi: 10.1016/j.cie.2009.09.003.

Li, X., Li, W., Gao, L., Zhang, C., & Shao, X. (2009). Multi-agent based integration of process planning and scheduling. In *2009 13th* International Conference on Computer Supported Cooperative Work in Design, Santiago, IEEE, pp. 215–220.

Li, X., Gao, L., Shao, X., Zhang, C., & Wang, C. (2010a). Mathematical modeling and evolutionary algorithm-based approach for integrated process planning and scheduling. *Computers & Operations Research*, 37(4), 656–667.

Li, X., Shao, X., Gao, L., & Qian, W. (2010b). An effective hybrid algorithm for integrated process planning and scheduling. *International Journal of Production Economics*, 126(2), 289–298.

Lian, K., Zhang, C., Gao, L., & Li, X. (2012). Integrated process planning and scheduling using an imperialist competitive algorithm. *International Journal of Production Research*, 50(15), 4326–4343.

Lihong, Q., & Shengping, L. (2012). An improved genetic algorithm for integrated process planning and scheduling. *International Journal of Advanced Manufacturing Technology* 58(5–8): 727–740.

Moon, C., Lee, Y. H., Jeong, C. S., & Yun, Y. S. (2008). Integrated process planning and scheduling in a supply chain. *Computers and Industrial Engineering*, 54, 1048–1061.

Moon C., & Seo, Y. (2005). Evolutionary algorithm for advanced process planning and scheduling in a multi-plant. *Computers and Industrial Engineering*, 48, 311–325.

Morad, N., & Zalzala, A.M.S. (1999). Genetic algorithms in integrated process planning and scheduling. *Journal of Intelligent Manufacturing*, 10(2), 169–179.

Moon, C., Jeong, C. S. & Lee, Y. H. (2003). Advanced planning and scheduling with outsourcing in manufacturing supply chain. *Computers and Industrial Engineering*, 43, 351–374.

Nasr, N., & Elsayed, E. A. (1990). Job shop scheduling with alternative machines. *International Journal of Production Research*, 28(9), 1595–1609.

Özgüven, C., Özbakır, L., & Yavuz, Y. (2010). Mathematical models for job-shop scheduling problems with routing and process plan flexibility. *Applied Mathematical Modelling*, 34(6), 1539–1548.

Park, B. J. & Choi, H. R. (2006). A genetic algorithm for integration of process planning and scheduling in a job shop. In *Australasian Joint Conference on Artificial Intelligence*, A. Sattar & B. H. Kang (Eds.). Springer, Berlin, Heidelberg, pp. 647–657.

Phanden, R. K., Jain, A., & Verma, R. (2013). An approach for integration of process planning and scheduling. *International Journal of Computer Integrated Manufacturing*, 26(4), 284–302.

Shao, X., Li, X., Gao, L., & Zhang, C. (2009). Integration of process planning and scheduling—a modified genetic algorithm-based approach. *Computers & Operations Research*, 36(6), 2082–2096.

Sundaram, R. M., & Fu, S. S. (1988). Process planning and scheduling. *Computers and Industrial Engineering*, 15(1–4), 296–307.

Wong, T. N., Leung, C. W., Mak, K. L., & Fung, R. Y. K. (2006). Integrated process planning and scheduling/rescheduling: An agent-based approach. *International Journal Production Research*, 44(18–19), 3627–3655.

Zhang, R., & Wu, C. (2010). A hybrid approach to large-scale job shop scheduling. *Applied intelligence*, 32(1), 47–59.

Index

A

Alternative process plans, 62, 162–163
  planning, 20
  fixed, 12, 162
  networks, 188–189
  routes, 28, 35, 96–100, 167, 170–171, 174–175
Ant colony optimization (ACO)
  algorithm, 11, 26, 27, 59, 39, 166, 186
Ant lion optimisation (ALO) algorithm, 27, 187, 190–194
Apparent tardiness cost (ATC), 30, 104, 166, 168, 176
  weighted ATC, 166, 168, 176
Artificial bee colony (ABC) algorithm, 117, 182
Assumption, 11, 12, 50, 52, 91, 96, 97, 121

B

Balanced level of machine utilisation, 190
Bill of materials, 96–97, 100, 102–103, 106, 109

C

Case study, 40, 84, 90, 92
Computer-aided process planning
  (CAPP), 4, 5, 14, 22, 29, 31, 33, 38, 39, 147, 162, 163, 185
Constant due-date (CON) 167, 181,
  see also Weighted, WCON
Customers, 12, 19, 162, 163
  elite, 164
  important, 163–164, 166–167, 181
  negotiations, 162

D

Decision-making criterion, 8

E

Dispatching rules, 166–168, 170, 176
Due-date, 162–163, 166
  assignment rules, 164, 166–67, 170, 174–175, 180–181
  common due-date assignment, 164–165
  due-date assignment (DDA), 162, 166
  due-date assignment function, 162
  due-date assignment process, 163
    (*see also* Stipulated due dates)
  external due date assignment, 162, 164, 181
  internal due-date assignment, 162, 164, 181
  long-due-date, 164 (*see also*
    Common due-date,
    Predetermined due-dates,
    Separate due windows and,
    Weighted due-dates)
  negotiation with customers, 162, 164, 181
  separate due-date assignments, 165
  weighed due-date assignment, 163–166
Dynamic Integrated Process Planning,
  Scheduling and Due-Date
  Assignment (DIPPSDDA) 165;
  see also IPPSDDA

Earliness, 162–163, 181
  weighted, 164
Earliest due-date (EDD), 168, 176
  weighted, 168
Earliest release date (ERD), 168, 176
  weighted, 168
Evolutionary strategies (ES), 165, 169, 171, 176, 177, 182
  hybrid RS/ES, 165, 174, 176–177, 182
Energy consumption, 12, 19, 28, 116–121, 125, 127, 129–131

F

Final planning phase, 31, 33, 38
Fitness function, 25, 52, 59, 73, 75, 77, 188,
        190, 194, 197–200, 202
Flow shop, 9, 10, 165, 167, 181
    two-machine flow shop, 165, 167

G

Gantt chart, 124, 129, 131, 197, 199, 200, 202
Genetic Algorithm (GA), 165–166, 169,
        171, 176, 177, 182
    GA Pseudo-Code, 172
    hybrid genetic search, 165
    hybrid RS/GA, 165, 174, 176–177, 182
Group technology, 4, 22

H

Horizontal flexibility, 210; *see also*
        Vertical flexibility
Hybrid moth flame optimisation (HMFO)
        algorithm, 121, 127, 129–132

I

Industry 4.0., 2, 13, 15
Integration of Process Planning,
        Scheduling and Due-Date
        Assignment (IPPSDDA), 162,
        165, 167, 169, 171, 174, 180–182
IPPS approaches, 20
    non-linear approach (NLA), 20, 21, 23,
        28, 39, 41, 42, 50, 53, 54, 71, 208
    closed-loop approach (CLA), 20, 28,
        29, 39, 41, 42, 50, 53, 54, 71
    distributed approaches (DAs), 20, 31,
        32, 36, 38, 39, 41, 42

J

Job shop, 9, 10, 165, 167, 181
    flexible, 96
    complex, 98
JOIN node, 188
Just-In-Time (JIT), 162, 164, 181
    process planning, 31

L

Longest operation time (LOT) 168, 176
    weighted, 168
Longest processing time (LPT) 168, 176
    weighted, 168
Linear ranking selection (LRS), 76

M

Machine flexibility, 187
Manufacturing flexibility, 13, 15, 20, 21,
        25, 26, 29, 31, 37, 187, 195, 198,
        208, 210, 211, 217, 220
    processing flexibility, 21, 25, 26, 187,
        209, 211, 213
    sequencing flexibility, 21, 25, 187, 209,
        211
    operational flexibility, 21, 25, 209
Mathematical model, 23, 27, 37, 40, 116,
        118, 119, 131, 187, 188, 201
Meta-heuristics, 169
    hybrid Meta-Heuristics, 174
    meta-heuristic algorithms, 166
    pure Meta-Heuristics, 169
Minimum Slack (MS), 168, 176
    weighted, 168
Multi Agent System (MAS), 26, 31, 112,
        186
Multi-objective hybrid moth flame
        evolutionary algorithm
        (MOO-HMFA), 120, 127
Multiple Process Plans (MPP), 20–27, 33,
        35, 50, 55–58, 69, 87–89, 211

N

Non-Polynomial hard (NP), 10, 20, 42,
        98, 110, 116, 120, 132, 162, 166,
        181, 186, 201
NSGA-II, 27, 118, 120, 121, 129–132
Number of operations (NOP) 167, 181
    weighted, 167
Number of Operation Plus Processing
        Time (NOPPT), 167
    weighted, 165, 167

O

Open shop, 9, 10; *see also* Mixed shop
Operation sequencing, 2, 24
Operation sheet, 3; *see also* planning
      sheet, route sheet, route plan,
      part program
OR network, 187
OR node, 188
Ordinary Solution (OS), 165, 176

P

Particle swarm optimisation (PSO),
      166, 172
   Hybrid RS/GA/PSO, 182
   Hybrid GA/MDPSO, 182
   Hybrid RS/MDPSO, 182
   MDPSO, 166, 172, 182
   MDPSO Pseudo-Code, 173
   Modified Discrete Particle Swarm
      Optimisation, 166
   Modified discrete PSO, 169
   Moment of inertia, 172
Performance function, 164
Preplanning phase, 31, 38
Problem Formulation, 9, 50
Probability Model, 137–138; *see also* for
      process plan selection, for the
      operation scheduling decision
Process planning, 2–5, 13, 14; *see also*
      Macro, Micro, Classification,
      Manual
   progressive, 31 (*see also* Collaborative,
      Concurrent and, Phased
      process planning approach)
   flexible process planning, 20 (*see also*
      Multi-process planning, Non-
      linear process planning)
Processing-time-plus-wait (PPW), 167, 181
   weighted, 167
Production Scheduling, 2, 6, 12, 14, 19,
      22, 23, 34, 42, 117, 142, 162–163,
      165–166
   scheduling approaches, 9, 96, 98, 99,
      166
   classification, 9

offline and online, 10
static and dynamic, 11
master production scheduling, 6
   (*see also* Detailed scheduling)

R

Random Search (RS), 165, 169–170,
      176–177, 182
Random-allowance due-dates (RDM),
      167, 175–176
Real-time-based process plan, 28
Repairing scheme, 80, 81
Restart scheme, 52, 82, 83
Rolling horizon, 96, 98–99, 107–109,
      111–112
Roulette wheel selection (RWS), 76, 193

S

Scheduling function, 162–163
Scheduling objectives, 6, 8, 28, 29
Scheduling with due window
      assignment (SWDWA), 163–165
Scheduling with due-date assignment
      (SWDDA), 162, 164–165, 180–181
Service in random order (SIRO), 168, 176
Slack (SLK), 165, 167, 181
   weighted, 166–167
   SLK due-date assignment, 165
Shop floor, 162, 163, 174, 178
   dynamic, 165
   utilization, 163
   work load, 162
   static, 165
Shortest Operation Time (SOT), 168, 176
   weighted, 168
Shortest Processing Time (SPT), 168, 176
   weighted, 168
Simulated Annealing (SA), 166, 169, 170,
      182
   algorithm, 169–170
   hybrid RS/SA, 174, 182
Simulation infrastructure, 31, 99,
      105–106, 112
   software, 39, 52, 54, 59, 62, 75, 82, 107
   horizon, 108, 110

Simulation-based scheduling algorithm, 24
    scheduling module, 34
    MGA, 39
    GA, 52–53, 58–59, 62, 71, 84, 92
    performance, 105
Single machine shop, 9, 10, 98, 108, 165, 167, 181
    multi-machine, 165, 167, 181
    parallel machine, 96, 165, 167, 181
Stochastic universal sampling (SUS), 76

T

Tabu Search, 170; *see also* Tabu list
    Semi-Tabu search (ST) 170, 182
    Hybrid RS/ST, 174, 182
    Hybrid semi-tabu search, 166
Tardiness, 162–163, 181
    weighted, 164
Temporal flexibility, 211; *see also* Short-term flexibility
Testbed problems, 41, 51, 84, 195, 207–11, 217, 218, 220
    small size IPPS testbed problems, 209, 218
    Extended size IPPS testbed problems, 210 (*see also* Small to medium size IPPS testbed problems)

Total work content (TWK), 167, 181; *see also* WTWK
    weighted, 167
Tool flexibility, 187
    completion time, 42, 119, 131, 136, 139
    processing cost, 119
    weighted tardiness (TWT), 96, 97
Tournament selection (TS), 76
Traditional manufacturing system, 162

U

Univariate marginal distribution, 137

V

Variable neighbourhood search (VNS), 96
    reduced, 101
    tailored, 103
Variant CAPP, 4, 5, 14; *see also* Generative, Semi-generative

W

Weighted scheduling, 163–166

For Product Safety Concerns and Information please contact our EU
representative GPSR@taylorandfrancis.com
Taylor & Francis Verlag GmbH, Kaufingerstraße 24, 80331 München, Germany